The Human Side of Science

ALSO BY ARTHUR W. WIGGINS

The Joy of Physics

C.1

The Human Side of Science

Edison and Tesla, Watson and Crick,
and Other Personal Stories
behind Science's Big Ideas

Arthur W. Wiggins and Charles M. Wynn Sr.

Cartoon Commentary by Sidney Harris

Prometheus Books
59 John Glenn Drive
Amherst, New York 14228

Published 2016 by Prometheus Books

Cartoon commentary by Sidney Harris
Cover design by Liz Scinta
Cover images:
Lightbulb (© Media Bakery)
Lise Meitner and Otto Hahn (from Wikimedia Commons, user Pieter Kuiper)
Watson and Crick (© A. Barrington Brown / Science Source)

Inquiries should be addressed to

Prometheus Books
59 John Glenn Drive
Amherst, New York 14228
VOICE: 716–691–0133
FAX: 716–691–0137
WWW.PROMETHEUSBOOKS.COM

20 19 18 17 16   5 4 3 2 1

Library of Congress Cataloging-in-Publication Data Pending

Printed in the United States of America

CONTENTS

PREFACE

J. Craig Venter (1946–). From Wikimedia Commons, user Margoz.

Francis Collins (1950–). From Wikimedia Commons, user Magnus Manske.

President William Clinton (1946–). From Wikimedia Commons, user Kmccoy.

Genome design is going to be a key part of the future. That's why we need fast, cheap, accurate DNA synthesis, so you can make a lot of iterations of something and test them.
> —J. Craig Venter, CEO,
> The Institute for Genomic Research (TIGR)[1]

One must dig deeply into opposing points of view in order to know whether your own position remains defensible. Iron sharpens iron.
> —Francis Collins, director,
> International Human Genome Sequencing Consortium[2]

Fix it. Make those guys work together.
> —William Jefferson Clinton, in F. Golden and
> M. D. Lemonick, "The Race Is Over," *Time*, July 23, 2003

Our previous efforts to make science attractive and accessible focused primarily on ideas and secondarily on the people who created the ideas. This book reverses the emphasis: people first, ideas second.

Should amount to the same thing, right?

Nope. Our research for this project revealed fascinating contentions among scientists pursuing similar subjects, helpful cooperation between some of them, and connections that led to new ideas. You will see examples of passion, greed, espionage, commitment, jealousy, sexism, impatience, obsession, animosity, envy, racism, and audacity as this parade of people pursue the puzzle of the natural world.

The more things change, the more they stay the same. Ideas change, but human interactions remain quite similar. The same atoms that Aristotle and Democritus argued about comprise the human genome that Venter and Collins raced to analyze.

In most chapters we've included a reference to "bonus" materials on a particular topic, in the form of videos, websites, or books. Please refer to the To Dig Deeper section for more interesting details.

We hope your view of science will be as enriched by this project as much as ours has been. Enjoy your journey through the history of science as you meet the fascinating people who made that history.

AWW
CMW
SH

SCIENCE'S EVALUATION SYSTEM

> Science is best defined as a careful, disciplined, logical search
> for knowledge about any and all aspects of the universe,
> obtained by examination of the best available evidence and
> always subject to correction and improvement upon discovery
> of better evidence. What's left is magic. And it doesn't work.
> —James Randi, *The Mask of Nostradamus*, 1993

This book needs an additional subtitle: "Contention, Cooperation, and Connection between a Wide Variety of Scientists throughout the World over the Last 2,500 Years That Resulted in Good, Bad, and Even Ugly Interactions among Scientists." That is a mouthful, but it gives you a far more complete picture of what you're about to read.

Our chronology begins in ancient Greece with conflict between the ideas of Democritus and Aristotle regarding the existence (or nonexistence) of atoms. Although Democritus and Aristotle never met in person, their ideas clearly influenced each other. It ends in the 1990s with contention between J. Craig Venter's Institute for Genomic Research and the Human Genome Consortium, headed by Francis Collins. This was good, in that both groups probably worked faster and more efficiently since there was a competition. But it was also bad in the sense that cooperation or collegial interactions between researchers became strained, and valuable insights were not shared in a timely fashion.

Isaac Newton's rivalry with Robert Hooke was bad in that the feud consumed valuable time that could have been spent more productively. But it was good in that the contention may have spurred both of them to accomplish more than they might have otherwise. Good and bad are actually in the eye of the beholder and may vary as time progresses. The real question here is the determination of BIG scientific ideas: they must be judged by objective standards rather than by subjective evaluations.

So before we start in on the juicy details of the lives and times of a wide variety of interesting people (almost four hundred in all) from around the world, let's take a look at the process that is intended to govern the evaluation of scientific ideas and determines which ideas are actually BIG. Ahh, now we're down to the real nitty-gritty: Science's evaluation process is the ultimate arbiter of scientific ideas.

Hang on; there's a good bit of detail here, but grasping the backbone of the scientific process is key to understanding what makes some ideas big and some not so big.

Let's start off by dealing with a giant misconception: To some people, science may seem like a huge, rambling house with many suites of rooms corresponding to the branches of science, all nice and neat and tidy.

Science Building. Used with permission from Sidney Harris.

THE ROAD TO REALITY

Science's process or evaluation system can be viewed in retrospect as a sequence of specific steps, often referred to as the scientific method:

> Observation: Certain specific happenings in physical reality are sensed.
> Hypothesis (H): A general idea of the nature of all such events is created.
> Prediction: Presuming the H to be true, some similar happening is forecast.
> Experiment: A test is conducted to see if the predicted outcome will occur. A new specific happening in physical reality is sensed as a result of the test.
>
> *match* => H is supported but not proved
> Compare Experiment with Prediction
> *doesn't match* => H must be modified

SCIENTIFIC OBSERVATION

Within the observation step, some specific, real occurrence is perceived by our senses with or without the aid of instrumentation. Viewing the sky with sharp eyes is a good start, but the telescope opened whole new vistas. While the natural sciences (e.g., physics, chemistry, biology, etc.) have a large number of identical subjects to observe (think carbon atoms), the human sciences (e.g., political science, sociology, economics, etc.) have a smaller number of distinctly different subjects (think human beings, even identical twins).

SCIENTIFIC HYPOTHESES

Human nature being what it is, information will be collected for just so long before the mind, in its search for order, begins to sift through this information and to construct patterns or develop explanations of these data. Viewing the stars is good, but noting that some seem to be fixed starts a whole new process. This is called the hypothesis step.

The reasoning that considers specific observations and constructs a general hypothesis is known as inductive logic, which involves making a generalization, and is the most precarious type of reasoning. Some people make an art form of jumping to conclusions, but within the context of the scientific method, that activity is restricted because succeeding steps help bring the hypothesis back to reality. Developing a hypothesis or a theory means about the same thing, but hypothesis is a more fancy-sounding word.

MATHEMATICS: A FOREIGN LANGUAGE

Often, the hypothesis is framed as a whole or in part in a language different from normal linguistic fare. That language is mathematics. Because mathematical skills require a great deal of effort to acquire, explaining scientific hypotheses to people not trained in mathematics requires translation into conversational language. Unfortunately, the meaning of the hypothesis often suffers in the process.

Mathematics is like vitamins—every science needs at least its minimum daily requirement. The type of mathematics used in the sciences varies according to need. Virtually all sciences make use of arithmetic to express their results in tangible form. In addition to arithmetic, algebra and geometry are used by all natural sciences. Physics and chemistry go further to include calculus as well as more esoteric forms of mathematics. Biology and the human sciences often employ statistics to characterize and evaluate their nonidentical populations in statistical terms, like averages (means) and variations from the norm (standard deviations). As the figure shows, some math is very advanced.

Square Root of Chicken. Used with permission from Sidney Harris.

SCIENTIFIC PREDICTION

Once a particular hypothesis is formulated, it is used to forecast some future event that will occur in a particular way if the hypothesis is true. This prediction is derived from the hypothesis using deductive logic. This form of logic starts from a true general statement and derives a true specific example from it. For example, Isaac Newton's second law of motion is expressed as $F = ma$, where F is force, m is mass, and a is acceleration. If $m = 3$ units and $a = 5$ units, then F should be 15 units. Carrying out this step is an ideal task for computers, which have deductive logic built into their programs. Newton's hypothesis was applied to a specific case, and a prediction was made that could be tested in reality.

SCIENTIFIC EXPERIMENTATION

Once the prediction is made, the next step is to perform an experiment to see if the prediction is supported by evidence. While this sequence is easy to state, in many cases it is extremely hard to accomplish. Intricate, expensive, labor-intensive scientific apparatuses have been constructed and operated by dedicated experimenters, and much valuable data has been collected. The natural sciences have the great advantage of being able to isolate the object of study (think test tubes), while the human sciences often have to contend with several variables being simultaneously filtered through the minds of people having their own agendas (think surveys in which participants complete questionnaires).

THE MATCH GAME

Once the experiment phase is complete, the result is compared with the prediction.

If the experiment *matches* the prediction, then the hypothesis is supported. Since the hypothesis is a general idea and the experimental results are specific results from reality, a specific favorable result can't necessarily *prove* a general hypothesis, it merely supports it. On the other hand, if the experimental result *doesn't match* the prediction, some aspect of the hypothesis is false. This feature of the scientific method is called falsifiability, and it places a stringent requirement on hypotheses. As Einstein said, "No amount of experimentation can prove me right, one experiment can prove me wrong."[1] Positive results only lend more support to the proposed hypothesis, but negative results can undermine it completely.

SCIENTIFIC RECYCLING

A hypothesis that is shown to be false in some way must be recycled. That is, it must be modified slightly, changed radically, or abandoned altogether. The judgment about how much to change can be an extremely difficult call. The recycled hypothesis will then have to work its way through the sequence again and hopefully survive the next prediction/experiment comparison.

AND NOW FOR THE FINE PRINT

Another facet of the scientific method that keeps the process on target is replication or repeatability. Any observer suitably trained and equipped should be able to repeat prior experiments and obtain comparable results. In other words, there's constant rechecking going on in science.

For example, a team of scientists at Berkeley Laboratory in California attempted to synthesize or produce a new element by bombarding lead targets with an intense beam of krypton ions (ions are atoms that have become charged by removing an electron) and then analyzing the resulting products. The Berkeley scientists announced their finding in 1999: one of the products was the synthesis of element 118 (118 is the number of protons in the atom's nucleus).

Synthesis of a new element is interesting news because of the new element's novelty. In this case, a favorable result would also support previous ideas about the stability of heavy elements like lead. Scientists at other laboratories in Germany, France, and Japan, however, were unable to duplicate the reported synthesis of element 118. An augmented Berkeley Laboratory team repeated the experiment, but they, too, failed to reproduce the earlier reported synthesis. After the Berkeley team reanalyzed the original experimental data using revised software codes and were unable to confirm the existence of element 118, they retracted their claim. *This refinement process shows that science's quest to understand reality is, and must be, a never-ending story.*

Scientific Method. Used with permission from Sidney Harris.

Another example of rechecking in science involves repeating the testing of a prediction. In February 2001, Brookhaven National Laboratory in Long Island, New York, reported an experimental result for a property known as the magnetic moment of the muon (a negatively charged particle similar to the electron but considerably more massive) that was slightly larger than the prediction from the Standard Model of particle physics. Because the Standard Model's predictions had been matched by experimental results to an extremely close tolerance for many other particle properties, there was a strong implication that this discrepancy in the magnetic moment of the muon indicated that the Standard Model was flawed.

The prediction was the result of a complex and lengthy calculation that had been done independently by groups in Japan and New York in 1995. In November 2001, the calculation of a muon's magnetic moment was repeated by physicists in France. The French physicists discovered an erroneous minus sign on one of the terms. They posted their results on the Internet. As a result, the Brookhaven group rechecked their own calculations, acknowledged the mistake, and published their corrected results. The net effect of this correction was to reduce the disagreement between the prediction and experimental results to an amount within the accuracy of the experiment. Although the Standard Model survived this challenge intact, it awaits and must withstand future challenges as science's never-ending search continues by testing to see if the hypothesis yields predictions that are matched by experimental evidence.

THE SCIENTIFIC METHOD IN ACTION

Let's follow an example of the scientific method at work, step by step:

OBSERVATION: J. J. Thomson, the director of the Cavendish Laboratories in England just before the turn of the twentieth century, observed what happened to a beam of light produced in a cathode-ray tube (forerunner of the modern TV picture tube). Since the beam (1) deflected toward positively charged electrical plates and (2) hit its target, producing individual flashes of light, it had to consist of negatively charged material since opposite charges attract, and the beam was attracted by positively charged plates. Since individual flashes of light were seen, this beam likely consisted of individual particles. These particles were named electrons by Irish professor George FitzGerald in his comments on Thomson's experiment.[2]

HYPOTHESIS: Since atoms are uncharged (neutral), and Thomson had found negatively charged particles within them, he deduced there must be some positive charge in atoms as well. In 1903, Thomson theorized that the positive charge was smeared throughout the whole atom, with the negatively charged electrons embedded inside the positive material. This depiction resembled a traditional British dessert and was therefore referred to as the Thomson Plum Pudding Model of the atom.

PREDICTION: Ernest Rutherford was an expert on positively charged particles known as *alpha* particles. He predicted that if these particles were shot at atoms consisting of the sparse and smeared-out positive charge of the Thomson Plum Pudding atom, it would be like shooting pool balls at a fog cloud. All particles ought to rip right through the smeared-out positive charge.

EXPERIMENT: In 1909, Hans Geiger and Ernest Marsden set up an apparatus to shoot alpha particles at an extremely thin sheet of gold atoms (gold was used because it could be made so thin). The results were somewhat different from what they expected. Although most of the alpha particles did go straight through, some alpha particles were deflected at large angles, and some even bounced back. Rutherford said, "It was almost as incredible as if you fired a fifteen-inch shell at a piece of tissue paper and it came back and hit you."[3]

RECYCLE: The Thomson Plum Pudding Model was replaced by the Rutherford Solar System Model, in which the positive charge was concentrated in a relatively tiny nucleus at the center of the atom (and thus could deflect a small number of alpha particles) and electrons (analogous to planets) that moved in circular orbits around the nucleus (analogous to the sun). (Recycling is often partial, in that some aspects of an earlier model are maintained, but it can be total, as we will see in later chapters.) Later in the twentieth century, as a result of subsequent predictions and experiments, the Rutherford Solar System Model of the atom was replaced by a series of other, more sophisticated models.

Whenever experimental evidence doesn't match the prediction of an existing hypothesis, it's time to recycle the hypothesis. Despite the popularity of an earlier idea, the celebrity status of a theory's proponents, the unattractiveness of a competing new theory, or the difficulty in understanding it, the bottom line is: *experimental evidence rules*.

THE NITTY-GRITTY: CONTENTION, COOPERATION, AND CONNECTION

Perhaps as a result of seeing scientific methodology presented in this way, some people believe science operates in a cut-and-dried fashion, with rational logic always prevailing. Those people are mistaken.

There are elephants in science's rooms. All the steps in the scientific method involve *people*. And you know what that means: The seemingly well-defined procedural steps of the scientific method are, when put into practice, actually fuzzy—and subject to the full range of human foibles.

We may be an imperfect lot, but we are very curious about our surroundings and can have several opinions about any one matter. Individually and
collectively, all sorts of *contention, cooperation, and connection*—and
even serendipity—take place. Often, we resist change and become contentious or even obnoxiously ugly. On the other hand, we might see immense
value in the ideas of our fellow humans and cooperate nicely. Additionally,
our seemingly vast world is actually much smaller than it appears. There
are connections among people that appear in unexpected places and con-
texts. This book aims to explore this complex territory, with an emphasis
on the people involved.

Elephant House. Used with permission from Sidney Harris.

CHAPTER 1

DEMOCRITUS AND ARISTOTLE PONDER THE EXISTENCE OF ATOMS

> It is as easy to count atomies as to resolve the propositions of a lover.
>
> —William Shakespeare, *As You Like It*,
> act 3, scene 2 (1599)

The idea that matter consists of atoms is pervasive in modern culture, even though no one has ever seen them directly. One of the best and most interesting pictures we've got was produced in 1989. It was obtained by a scanning tunneling microscope at the IBM Almaden Research Center in San Jose, California. A beam of xenon atoms was shot at a chilled nickel crystal, and the atoms that stuck to the crystal were manipulated by the microscope's probe into the pattern shown below:

"IBM" spelled out in 35 atoms. Courtesy of IBM.

SLICING AND DICING MATTER

The reality of atoms answers a fundamental question that is simple enough to frame: If any material object is cut into smaller and smaller pieces, would there be some limit, after which no further cuts could be made, or would the material be infinitely cuttable?

Ancient Greeks. Used with permission from Sidney Harris.

To get to the heart of this question, we need to go back, back, way back. How far? As many wags have suggested, "The ancient Greeks thought of everything first." Atoms are a case in point.

Leucippus (fifth century BCE). From Wikimedia Commons, user Jean-Jacques MILAN.

Democritus (460 BCE–370 BCE). By Antoine Coypel (1661–1722). From Wikimedia Commons, user Fæ.

DEMOCRITUS: MATTER IS DISCONTINUOUS

Democritus (460 BCE—370 BCE) was born in the far north of Greece, in Abdera. He was taught by Leucippus, and many of his and Leucippus's ideas are so interwoven they are difficult to separate. As a young man, Democritus traveled extensively, spending a great deal of his inheritance from a wealthy father. He visited Asia, Ethiopia, India, and spent a long time in Egypt. Democritus's writings have not survived intact, so only secondary sources exist to help convey his ideas.

> I am the most travelled of all my contemporaries; I have extended my field of enquiry wider than anyone else; I have seen more countries and climes and have heard more speeches of learned men.[1]

Although he was virtually unknown in Athens (he went there once to visit Anaxagoras and was ignored), his students and neighbors in Abdera regarded him highly because he was invariably cheerful and ready to see the humorous side of life. He was referred to as the "Laughing Philosopher." He and Leucippus were materialists in that they looked for mechanistic explanations of phenomena, rather than deeper causes. They thought perception through the senses was so subjective that it was unreliable, so they discounted observations as relevant evidence.

Democritus (and, by extension, Leucippus) answered the fundamental question about divisibility of matter by postulating the existence of an uncuttable unit called an atom (Greek *a* = "not," *tomos* = "to cut") as the fundamental unit of matter.

> The first principles of the universe are atoms and empty space; everything else is merely thought to exist.[2]

Aristotle (384 BCE–322 BCE). By Raphael (1483–1520). From Wikimedia Commons, user Dencey~commonswiki.

ARISTOTLE: MATTER IS CONTINUOUS

Aristotle (384–322 BCE) was born in the small city of Stagira in northeastern Greece, and he studied at Plato's Academy in Athens, where he started at about age seventeen. Plato (427–347 BCE), in turn, was a student of Socrates (469–399 BCE). After Plato's death, Aristotle left Athens for about twelve years. During that time, he tutored young people who went on to become kings, including Alexander (the Great). Upon his return to Athens in about 335 BCE, Aristotle established a school called the Lyceum.

Athens was the cultural hub of Greece and featured a tradition of discussion of all issues great and small. Aristotle and his students walked as they talked, earning them the title the "Peripatetic Philosophers." Although Aristotle agreed with Plato about many things, he and Plato also had significant differences of opinion. For example, Plato's reasoning was deductive in that it proceeded from general principles to specific instances. For example, all chairs have four legs, therefore this chair has four legs. On the other hand, Aristotle's reasoning was inductive in that it broadened from the specific to the general. For example, I have seen five tables, all of which have four legs. Therefore, all tables have four legs. Aristotle's emphasis on specifics led him to make systematic observations about living things, especially marine life, which set biology on a sound footing that lasted for centuries.

Aristotle invented and systematized logic as a tool for scientific inquiry. The overall scientific process used by Aristotle was one of careful observation, followed by logical reasoning about causes. Despite his extraordinarily wide-ranging scholarship that laid the groundwork for many intellectual pursuits during his own time and long thereafter, even a man as perceptive as Aristotle surprisingly made a few errors. His analyses of motion in general and planetary motion in particular were both flawed, leading to later difficulties, as we'll see in subsequent chapters. One particular error seems quite blatant, namely:

> Males have more teeth than females in the case of men, sheep, goats, and swine.[3]

With his emphasis on careful observation, many have wondered why Aristotle didn't simply count the teeth of both sexes and make a proper comparison.

More than two hundred treatises are credited to Aristotle. The thirty-one surviving works are suspected to be rough lecture notes, perhaps even taken by students and not intended for publication. Most of these were lost for a while but were later translated into Arabic and analyzed extensively by Muslim scholars. Some think Aristotle was the last person who knew everything that could be known at the time he lived.

Aristotle's brain. Used with permission from Sidney Harris.

When analyzing the question about matter's fundamental constituents, Aristotle comes down squarely against there being any smallest unit:

> Neither is there a smallest part of what is small, but there is always a smaller (for it is impossible that what is should cease to be). Likewise there is always something larger than what is large.[4]

Aristotle thought matter was made from the four elements suggested by one of his predecessors, named Empedocles, who identified them as fire, air, earth, and water. Aristotle added one more element, æther, the material of heavenly bodies.

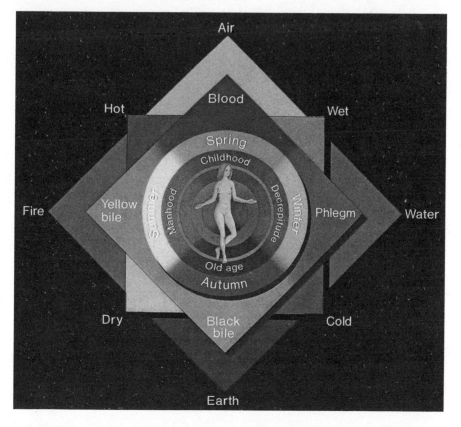

Aristotle's Elements. This file comes from Wellcome Images, a website operated by Wellcome Trust, a global charitable foundation based in the United Kingdom. From Wikimedia Commons, user Fæ.

DEMOCRITUS VERSUS ARISTOTLE

There is no record of Democritus and Aristotle ever having met, however, Aristotle knew of Democritus and his ideas and argued forcefully against them. One of his treatises was titled "On Democritus," but, unfortunately, only commentary survives.

Here is an example of Aristotle's argument against the atomic theory of Democritus:

> He (Democritus) ascribes the genesis and the separation opposed to it not only to animals but also to plants and to worlds, and comprehensively to all sensible bodies. If, then, genesis is combination of atoms, and destruction separation of them, then even according to Democritus "genesis" must be change of quality. Indeed, Empedocles, too, says that that which comes into being is not the same, except in kind, with that which has perished, and yet Alexander says that Empedocles assumes the existence of change of quality, not of coming into being. Are we then to say that the All is composed of indivisible substances? Some thinkers did, in point of fact, give way to both arguments. To the argument that all things are one if being means one thing, they conceded that not-being is; to that from bisection, they yielded by positing atomic magnitudes. But obviously it is not true that if being means one thing, and cannot at the same time mean the contradictory of this, there will be nothing which is not, for even if what is not cannot be without qualification, there is no reason why it should not be a particular not-being. To say that all things will be one, if there is nothing besides Being itself, is absurd. For who understands "being itself" to be anything but a particular substance? But if this is so, there is nothing to prevent there being many beings, as has been said. It is, then, clearly impossible for Being to be one in this sense.[5]

Although this may seem befuddling to us, perhaps it made more sense in Aristotle's day.

Aristotle's teacher, Plato, was said to be even more extreme in his opposition to Democritus. Although Plato's writings never mentioned Democritus by name, Aristotle's pupil, Aristoxenus, wrote that Plato wanted to burn Democritus's books, but they were too widely distributed for his idea to succeed. Ironically, the major reason we know about Democritus's ideas is because of their mention in Aristotle's students' writings. Since Aristotle taught in Athens and had many students and followers, his ideas would have won any popularity contest with those of Democritus.

Bonus Material: Aristotle/Democritus Internet interview. See To Dig Deeper for details.

AND THE WINNER IS . . .

Actually, neither Democritus nor Aristotle was completely correct.

In the modern view, matter is made of small units, usually molecules. Molecules are, in turn, made from atoms, but these atoms are not indivisible, as Democritus thought. Atoms have an outer cloud of negatively charged electrons moving around a positively charged nucleus. The nucleus contains positively charged protons and neutral neutrons. Neither of these, however, are fundamental particles. They are in turn made of even smaller particles called quarks. And quarks are made of . . . well, we don't know yet. Further, unlike the situation in ancient Greece, today's scientific ideas require the support of experimental evidence, which can get quite involved, to say nothing of the expense of conducting experiments to obtain such evidence.

Ideas were paramount in the Greek times, but there was no system of checks and balances in science the way it operates in the modern world. The next chapter takes us on a sometimes painful journey toward the modern system.

Philosopher-scientists. Used with permission from Sidney Harris.

CHAPTER 2

ARISTOTLE, ARISTARCHUS, COPERNICUS, AND GALILEO SEEK TO DETERMINE EARTH'S PLACE IN THE COSMOS

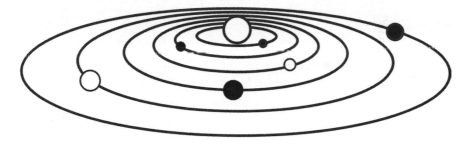

Solar system. Used with permission from Eugene Mann.

It is far better to grasp the universe as it really is than to persist in delusion, however satisfying and reassuring.

—Carl Sagan[1]

To grasp the real significance of the orbits in which planets circulate around the sun, we've got to go way back again.

Turn off your computers, your Internet connection, iPhone, iMacs, iPad, iWatches, iWhatevers, and go outdoors. It would be best to get as far from city lights as possible, on an evening with little moonlight and minimal clouds.

Look up.

What do you see?

Stars, of course.

Nothing new here. People have seen the lights that we call stars in the sky since before history was recorded.

Let's begin our story at about 3000 BCE with a look at the world of the Minoan sailors from Crete. The Minoans dominated trade in the Aegean and Mediterranean Seas. They had sturdy ships that used both oars and sails. Their ships were so good that some speculate that they could have sailed on the ocean, and possibly even ventured as far as the New World. The Minoans had a fabulous civilization for a while, but their culture was cut short by the eruption of the volcano Thea on the island of Santorini sometime between 1500 and 1600 BCE. The resulting tsunami devastated the Minoans, who were subsequently conquered by mainland Greeks. The beauty and sophistication of the Minoan cities and the sudden disappearance of their entire civilization led to a number of fables, including one that identified their region as the lost continent of Atlantis.

BACK TO THE STARS

Astute sailors had recognized the usefulness of stars in assisting navigation in the dark waters of the Aegean, and farmers had tracked the position of the sun to tell them when to plant and harvest in the appropriate season. But did these stars and constellations of stars have more than utilitarian value?

DEEPER MEANING

The Minoan sailors brought more than goods back and forth across the Mediterranean. They learned older ideas about stars and constellations from the Babylonians and Sumerians who lived in the Middle East, brought them across the Mediterranean, and in turn taught them to the Egyptians. Joining stars with lines like a connect-the-dots puzzle yielded interesting figures in the sky. For example, most everyone has seen the three stars in a row that make up the belt of Orion, who was a hunter figure in Babylonian lore. Interestingly, rather than focus on the lights, ancient South American astronomers (Incas) saw patterns in the dark areas of the sky such as the constellation Llama, in which the stars Alpha Centauri and Beta Centauri form its eyes, while dark spaces make up the body. Initially named for things they seemed to resemble, constellations later became associated with gods, and sometimes were actually thought of as being gods. Great significance was especially given to one group of constellations through which the sun and moon traced their narrow path. This plane is called the ecliptic, and it makes a complete 360-degree circle in the sky. The constellations along the ecliptic are called the zodiac constellations. The zodiac formed the basis for astrology, which was thought by many to be influential in determining the actions of people.

Besides the fixed stars, whose positions were always reliable, another group of lights in the sky was observed to move around a lot. The Greeks called these other lights moving lights planets, meaning wanderers. Such planets were eventually named for Roman gods and, in the minds of many people, assumed heavy roles in determining the course of the future. And then there were objects that suddenly appeared and then disappeared. These were called comets. Many believed comets must signal imminent changes in human affairs, unlike the stars and planets, which followed predictable paths. Eventually, interpretation of events in the heavens and their impact on humans fell under the jurisdiction of the priests and astrologers.

The history of star-gazing is fascinating stuff, but let's move on by fast-forwarding many centuries. We need to have another look at the work of one of the intellectual giants we met in the last chapter. Who might it be that would have the intellectual breadth to integrate the best ideas from past civilizations, some from his immediate teacher, some from his colleagues and, in addition, to add some touches of his own?

You got it. Aristotle.

Here is Aristotle's model of the structure of the whole universe:

Aristotle's universe. By Peter Apien, *Cosmographia* (1524).
From Wikimedia Commons, user Fastfission~commonswiki.

- Earth is a sphere, located at the very center of the entire universe.
- The moon, the sun, and the planets are all arranged on larger crystalline spheres surrounding Earth (around fifty of them).
- These spheres are made of a perfect material, called æther, and rotate around the earth at uniform speeds.
- The outermost of these spheres houses the fixed stars, and its rotation is caused by the "Prime Mover."

(Aristotle is vague about the Prime Mover's details.)

Since the earth is at its center, Aristotle's model as well as other similar ones is called a geocentric model.

TIME FOR SOME CONTENTION

Aristotle had so many ideas about such a wide variety of topics that he was a big target for disagreement about the validity of his concepts. A later member of his Peripatetic School, Aristarchus (310 BCE–230 BCE), observed the moon carefully and used the mathematics of trigonometry to estimate the size of the sun.

Aristarchus (310 BCE–230 BCE). From Wikimedia Commons, user Christian1985.

According to Aristarchus's measurements and calculations, the sun was substantially larger than the earth, so he thought it made more sense for the smaller earth to orbit the larger sun. Aristarchus's original work didn't survive, so we have only comments about his theory from Archimedes, twenty-five years his junior:

> His hypotheses are that the fixed stars and the Sun remain unmoved, that the Earth revolves about the Sun on the circumference of a circle, the Sun lying in the middle of the Floor.[2]

This model places the sun at the universe's center and so is referred to as a heliocentric model.

Since Aristarchus and Aristotle didn't overlap time-wise, they never had direct interchange about their differences. If a poll had been taken, Aristotle probably would have prevailed, but this was way before physical evidence became science's determining factor. Nevertheless, observations continued, and all astronomical data became more accurate. New information necessitated changes in Aristotle's model, especially because astrological predictions were often mistaken, and these errors were attributed to faulty data about planetary positions.

Ptolemy (90 CE–168 CE). From Wikimedia Commons, user Ineuw

Claudius Ptolemy (90 CE–168 CE) systematized, clarified, and augmented Aristotle's model in his work titled *Almagest*, published in 150 CE. Ptolemy was probably ethnically Greek, lived in Alexandria, Egypt, as a Roman citizen, and wrote in Greek. The *Almagest* included a star catalog that listed forty-eight constellations and *Handy Tables* that tabulated all the data needed to compute the positions of the sun, moon, and planets; the rising and setting of the stars; and eclipses of the sun and moon. Ptolemy's model was geocentric. It contained a series of nested spheres quite similar to Aristotle's. The major difference between Ptolemy's and Aristotle's models was that in Ptolemy's, Earth was displaced slightly from the center (called eccentric), and planetary orbits were circles whose centers were situated on the larger circles that orbited the earth. These are called epicycles. This variation was required to account for the occasional counterintuitive west-to-east motion of planets (called retrograde), which appeared to violate the normal east-to-west motion of heavenly bodies.

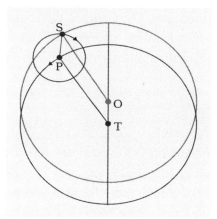

Epicycles. From Wikimedia Commons, user Hikin1987.

Ptolemy's *Almagest* was preserved in Arabic manuscripts. It was studied extensively over several hundred years after the collapse of the Western Roman civilization (400s CE). During the Middle Ages (474 CE–1500 CE), astronomy took a backseat to the culture's more pressing political and religious concerns, so Ptolemy's work stood supreme, since few changes were proposed to anything astronomical.

ENTER THE PHILOSOPHERS, THEOLOGIANS, AND TRANSLATORS

During the twelfth and thirteenth centuries, as the intellectual reawakening of the Renaissance began, interest in philosophy became much stronger. Aristotle's ideas were still significant but were accessible only through commentaries on translations. Lecture notes (not finished manuscripts) originally written in Greek had been translated into Syriac and Arabic, and were commented upon by the Persian philosopher Avicenna (Ibn Sina) and the Andalusian philosopher Averroes (Ibn Rushd). They each attempted to reconcile Aristotle's philosophy with Islamic teachings. The Syriac/Arabic translations were then retranslated into Latin and Hebrew, and the Spanish Sephardic Jewish philosopher Moses Maimonides (Moshe ben Maimon) commented on their relationship to Judaic teachings. As you might imagine, Latin translations of the work of Islamic and Judaic philosophers and comparisons to their religion's teaching led to much confusion and contention among Westerner Europeans. The Catholic theologians were particularly concerned about whether conflicts between Catholic teachings and Aristotelean ideas were the result of misinterpretations or mistranslations.

BREAKTHROUGH

Greek manuscripts of Aristotle's lecture notes were discovered. The excellent translator William of Moerbeke was sent by the Dominican theologian Albertus Magnus to Byzantium (now Istanbul) to translate them. Because of language similarities, translation from Greek directly to Latin generated far fewer translation errors than the multiple translations into the structurally dissimilar Syriac/Arabic, and cleaner translations were thus possible. Once the translations were complete, another student of Albertus Magnus, Thomas Aquinas, set out on an extremely ambitious project: to reconcile Aristotle's philosophical works with Catholic teaching.

Thomas Aquinas was remarkably successful, elevating much of Aristotle's thought almost to a par with scripture. Along the way, this included Aristotle's ideas about the solar system, as updated by Ptolemy. Thus, we arrive at 1500 CE with Aristotle's astronomy, at the heart of Catholic teaching.

CONTENTION, HERE WE COME

Next, the Polish astronomer Nicolaus Copernicus (1473–1543) rediscovered Aristarchus's work, then made measurements of his own that convinced him that the sun is the center of the universe, with all planetary spheres and the earth rotating about the sun.

Nicolaus Copernicus (1473–1543). From Wikimedia Commons, user Ineuw.

His ideas, released initially only in partial form, were not well received at first by Protestant theologians, and then by Catholic ones. This was because they didn't fit Ptolemy's model, which, thanks to Thomas Aquinas's synthesis, had (almost) the force of scripture. Hoping to avoid controversy, Copernicus delayed publication of his complete work until he was on his deathbed.

MORE CONTROVERSY

Tycho Brahe (1546–1601). From Wikimedia Commons, user Szajci.

Tycho Brahe was a larger-than-life fellow. He was born in Denmark and lived from 1546 to 1601. His parents had promised to give one of their male children to Tycho's uncle Jørgen and aunt Inger, who were childless, and so he was raised by his uncle and aunt from the age of two. They were quite wealthy and made sure young Tycho had a first-class education. Although Uncle Jørgen wanted Tycho to study law, his initial studies at the University of Copenhagen were quite general and included astronomy. Tycho witnessed an eclipse of the sun when he was fourteen, and it impressed him mightily. At age twenty, Tycho fought a duel with a cousin after a disagreement over a mathematical formula. The duel took place in the dark, and the bridge of Tycho's nose was sliced off. He wore a prosthetic nose made of brass for the rest of his life.[3]

In 1565, Denmark's King Frederick II had a disaster that greatly influenced Tycho's development. The king was thrown off his horse into a river. Jørgen jumped into the water and saved the king from drowning. Unfortunately, Uncle Jørgen contracted pneumonia and died. Several years later, Tycho's real father also died. With help from another uncle, Steen Bille, Tycho built an observatory. After noting the sudden appearance of a new star in the constellation Cassiopeia, which he called a nova (it was actually a supernova), Tycho realized the crucial importance of accurate measurements. This impelled him to make astronomy his life's work. Realizing the existence of a new star contradicted Aristotle's view of the heavens as unchanging, Tycho began writing and touring to discuss the deficiencies of the old ideas. King Frederick II recognized Tycho's talents and gave him the island of Hven to build an observatory. This observatory was called Uraniborg and was completed before Tycho turned thirty. Uraniborg featured the world's largest and most precise sextant and was supported by the king with 5 percent of Denmark's budget. Tycho Brahe produced the best astronomical data of his time, both in accuracy and amount. He was known as the last and best naked-eye (non-telescopic) astronomer.[4]

His conception of the universe was a curious hybrid of geocentrism and heliocentrism. As he saw it, all planets except the earth orbited the sun; the sun, the moon, and fixed stars orbited the earth just as Aristotle had thought.

THAT ALL SOUNDS REALLY GOOD, BUT WAS THERE ANYTHING BAD ABOUT TYCHO?

Tycho fell in love with Kirsten Hansen, daughter of a Lutheran minister. Kirsten was a commoner, and Tycho a nobleman, so their marriage was of the common-law variety, and their children could not inherit Tycho's land holdings.

Because of his land grant from the king, Tycho's neighbors on the island of Hven were required to pay him rent. He would go after them legally to collect, earning their ill will. Tycho was known to keep a pet elk that liked to attend parties and drink beer.[5] When Frederick II died, his son became the new king, Christian IV. Christian summoned Brahe to discuss his budgetary support, but Brahe put him off for months, pleading that he was too busy. When Christian cut off Brahe's funding, Brahe traveled to Prague to become the royal astronomer for the Holy Roman Emperor Rudolf II. Some suggest that the new king's displeasure might have been related to a possible affair between Tycho and the queen, but there is little evidence of that.

Brahe was difficult to work for and went through many assistants, with the last one being the most interesting: Johannes Kepler.

STRANGE CONNECTION OUT OF LEFT FIELD

In Tycho's family tree were cousins named Rosencrantz and Guilden-stern, whose names show up in Shakespeare's *Hamlet*. Also, when James VI of Scotland traveled to Denmark in 1590 to collect his bride, Anne of Denmark, a storm forced his party to spend some time as Tycho's guest on Hven. James VI succeeded Queen Elizabeth of England as James I in 1603 and became Shakespeare's patron.

MORE CONNECTION (AND POSSIBLY CONTENTION)

Johannes Kepler (1571–1630) was a German mathematician, astronomer, and astrologer. As a child, he witnessed a lunar eclipse and saw a comet, both of which impressed him greatly and inspired an abiding love for astronomy. His family fortune was in decline, so he helped make ends meet by casting horoscopes while still in school at the University of Tubingen. Although he was personally skeptical about much of astrology, he practiced it with some disclaimers. As a bright child of poor parents, he successfully passed through the school system, ending up at the University of Tubingen. It was at Tubingen that he became convinced of the reality of the heliocentric Copernican system. Only months before graduation, he was instead sent to be a teacher at a seminary school in Graz, Austria.

While teaching, he had an epiphany about planetary orbits. He became convinced that if the Platonic solids were nested within each other, and spheres inscribed and circumscribed each solid, the spheres would describe exactly the orbital radius of each of the six known planets.

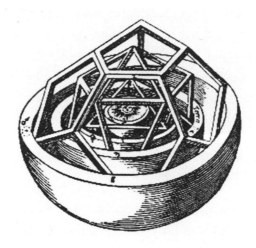

Kepler's solids. By Johannes Kepler (1571–1630), *Mysterium Cosmographicum* (1596). From Wikimedia Commons, user Hellisp.

As long as the solids were placed in the order: octahedron, icosahedron, dodecahedron, tetrahedron, cube, the spheres would describe the planetary orbits closely but not precisely within the accuracy of current measurements. Kepler published this idea in *Mysterium Cosmographicum* (*The Cosmic Mystery*). From his viewpoint, Kepler thought he had deduced God's geometrical plan for the solar system. His work was criticized by Tycho, who had much more accurate data on planetary orbits than was available to Kepler. He corresponded with Tycho on this. After more letters and breakdowns in negotiations, Kepler finally came to work as Tycho's assistant in January 1600. Although their relationship was stormy, Tycho began to trust Kepler, gave him access to more data, and introduced him to Rudolf II, for whom he was compiling astronomical data to be issued as the *Rudolfine Tables* (star catalogs and planetary orbital data).

Suddenly, in late 1601, Tycho died under mysterious circumstances a few days after a heavy drinking session. Soon after his death, his heirs sued

Kepler to get Tycho's data for their own. But delaying tactics by Kepler allowed him to hang onto the data he needed. Kepler became the new Imperial Mathematician and advisor to Rudolf II. Details of the orbit of Mars in particular convinced Kepler that his nested-solids scheme was insufficient. Looking for another way to solve the orbit problem, Kepler tried various ovoid-shaped orbits without success and finally used well-defined elliptical orbits, which worked beautifully. Kepler abandoned circular orbits and substituted ellipses because circles were inaccurate by eight minutes of arc in Tycho's accurate measurements of Mars's orbit. Kepler knew that Tycho's data was so accurate that it must all be matched precisely by whatever mathematical form the orbits took.

> Now, because they could not be disregarded, these eight minutes alone will lead us along a path to the reform of the whole of Astronomy.[6]

Three hundred years after his death, Tycho's body was exhumed because of suspicion about his cause of death, and mustache hairs were found to contain mercury, but not enough to justify speculation about the possibility of mercury poisoning.

CONTEMPORANEOUS WITH TYCHO AND KEPLER WAS GALILEO GALILEI

Galileo Galilei (1564–1642) was born in Pisa, Italy. His father was an accomplished lutenist and musical theorist, and he contributed to the theory of musical dissonance. Money was tight, so Galileo was encouraged to study medicine. While at the University of Pisa, Galileo noticed that chandeliers that swung because of air currents seemed to have the same period of oscillation (time to complete one cycle) whether the swing was wide or narrow. He timed the swings by his pulse. Although he had been purposely kept from studying mathematics, he finally convinced his reluctant father to allow him to study natural philosophy. After inventing the thermoscope (a forerunner of the thermometer) and a hydrostatic balance (an improvement on Archimedes's eureka moment because it was more accurate), his

Galileo Galilei (1564–1642). By Caravaggio (1571–1610). From Wikipedia Commons, user Sir Gawain.

ideas became noticed, and he was given a teaching position first at the University of Pisa, and then later at the University of Padua.

While at Padua, a former student of Galileo's wrote him about an invention he had seen in the Netherlands. Hans Lippershey had fitted two glass lenses into a tube, and when you looked through it, distant objects appeared closer. This was called a spyglass. It magnified images by a factor of three. Galileo was quite interested and had his instrument maker, Marcantonio Mazzoletti, make a better one, which magnified eight times. He sold this spyglass to the Venetian Senate, so it could be used to see who was entering the harbor. The Senate awarded Galileo with a lifetime faculty tenure at the University of Padua. Next, Galileo commissioned Mazzoletti to make a twenty-power model, which Greek theologian Giovanni Demisiani called a telescope. Galileo pointed it skyward. He found things never before seen with the naked eye: craters on the moon, spots on the sun, and satellites orbiting around Jupiter. Immediately, he noted the disagreement of these observations with the ideas of Aristotle, whose theory said heavenly bodies are perfect and unchangeable, and everything orbits the earth.

Another difference between Galileo's and Aristotle's ideas had to do with Aristotle's assertion that heavy bodies fall faster than light ones. Some have suggested that Galileo dropped heavy and light bodies from the Leaning Tower of Pisa in the presence of various authorities. When the bodies landed at their feet simultaneously the authorities were forced to choose between the written words of Aristotle and the evidence of their own senses, although there is no historical evidence to support this actual demonstration.

Galileo and the Leaning Tower of Pisa. Used with permission from Sidney Harris.

MR. CONGENIALITY?

In 1610, Kepler requested a telescope, but Galileo told him he had no extras and was too busy to even discuss it. A little competitive drive, there?[7]

Galileo carried on a dispute with the Jesuit priest Orazio Grassi that began about the nature of comets but spilled over into disagreement about the nature of science in general. The quarrel was carried on using pseud-onyms, but the identities of the combatants became known. The papers included insulting comments and, as a result, made numerous enemies on both sides.

Galileo also had a dispute with the Jesuit priest Christoph Scheiner regarding sunspots. The controversy took nasty turns with both combatants accusing the other of plagiarism.

Further, Galileo wrote in Italian, the language of the common people, rather than Latin, the usual academic language, so his papers enjoyed wider distribution and attention than those of his academic colleagues.

The relationship between the Roman Catholic Church and science was becoming ever more uneasy. The Catholic Church's Inquisition to root out heretics and blasphemers was extremely powerful. The Italian scientist/ mystic Giordano Bruno was burned at the stake for wide-ranging heresy, including some scientific ideas in 1600. Galileo's support for the heliocentric theory of Copernicus placed him under suspicion by the Inquisition, whose members wanted Galileo to state publicly that his theories about a sun-centered universe with imperfect celestial bodies were merely mathematical conveniences and not representative of reality.

But Galileo had an ace up his sleeve. The new pope, Urban VIII, was Maffeo Barberini, whom Galileo had tutored as a youth. The pope requested that Galileo explain both sides of the geocentric/heliocentric controversy and include the pope's own views prominently. Galileo obliged in his

famous work *Dialogue concerning the Two Chief World Systems*.[8] In it, he portrays a discussion between fictitious characters, with the official Catholic position being argued (ineffectively) by a character named Simplicio. This name may have honored a similar character in Aristotle's work, the philosopher Simplicius, but it is disturbingly close to the Italian word for *simpleton*. Galileo's enemies and those involved in court intrigue and problems of state convinced the pope that he needed to show strength, so Galileo had to be

Pope Urban VIII (1568–1644). By Caravaggio (1571–1610). From Wikimedia Commons, user Sir Gawain.

silenced. Galileo was convicted of being "vehemently suspect of heresy" and was sentenced to imprisonment. Even as he was being sentenced, Galileo was still rebellious. Supposedly, he muttered, "Eppur si muove," ("and yet, it moves"). The prison term was commuted to house arrest, so Galileo was unable to travel for the last nine years of his life.[9]

Bonus Material: Tycho/Galileo Internet interview. See To Dig Deeper for details.

The uneasy science/religion relationship eventually turned into a divorce, with science's final test becoming the existence of physical evidence, while the Church demanded faith regardless of physical evidence. This development is often referred to as the Scientific Revolution. Galileo certainly wasn't single-handedly responsible for this giant change, but he did a lot to install mathematics as the language of science and experimental evidence as the chief decider.

In the next chapter, the role of experimental evidence in science is so strong it is almost taken for granted.

CHAPTER 3

ISAAC NEWTON, ROBERT HOOKE, AND GOTTFRIED LEIBNIZ ARGUE ABOUT MOTION AND CALCULUS

> The astronomers said, "Give us matter and a little motion and we will construct the universe. It is not enough that we should have matter, we must also have a single impulse, one shove to launch the mass and generate the harmony of the centrifugal and centripetal forces." . . . There is no end to the consequences of the act. That famous aboriginal push propagates itself through all the balls of the system, and through every atom of every ball.
> —Ralph Waldo Emerson, *Works*, 1883, p. 122.

Moving bodies, heavenly or not, flummoxed the best minds for centuries. But all it took to get things on the right track was one genius, Isaac Newton, and a public health crisis (the Great Plague) that killed 25 percent of London's population.

Isaac Newton (1642–1727).
Used with permission from Sidney Harris.

NEWTON: THE EARLY YEARS

The year 1642 was a phenomenal year for science. At its beginning,

Galileo died, and at its end, Isaac Newton came into existence—but just barely. According to his mother, Hannah Ayscough (pronounced *askew*) Newton, premature baby Isaac would have fit into "a quart mug."[1] But Isaac's health was only one of several worries for his mother. Isaac's father, also named Isaac, had died three months earlier. So here was an English widow, nursing a premature infant and trying to run a sizable farm at the beginning of the violent English Civil War (1642–1651). Fortunately, things got better for her fairly quickly. Within two years after Isaac's birth, she received a marriage proposal from a sixty-three-year-old childless widower clergyman, Barnabas Smith. It was an offer she couldn't refuse. The wealthy Pastor Smith promised to take care of young Isaac's future by providing him with income from a property, which he would acquire when he reached maturity.

In the meantime, Isaac was to be cared for at the family home, Woolsthorpe, by his Ayscough grandparents. They had come to run the farm while Isaac's mother moved in with Pastor Smith, just two miles away. Newton's mother bore Smith three children in six years. Smith died in 1653, as did Newton's grandfather, James Ayscough. By now a wealthy widow, Hannah Ayscough Newton Smith moved back to Woolsthorpe to live with her younger children (Mary, Benjamin, and Hannah), her mother, and Isaac, now ten years old.[2]

THE UNWILLING FARMER

In less than a year, Isaac was sent off to the King's School in Grantham, five miles away. But he didn't live at home; he boarded with a neighboring family, the Clarks. There are mixed reports about his success at school, but he clearly enjoyed his leisure time in which he built toys, windmills, clocks, and furniture. Eventually, his mother decided Isaac needed to learn how to run the family farm, so she brought him home and assigned a trusted servant the task of making a farmer out of him. This enterprise failed because Isaac preferred to read books or build clocks rather than run a farm. His uncle William Ayscough finally persuaded Hannah to let Isaac attend his alma mater, so Isaac enrolled in Trinity College, Cambridge, in 1661.

At Cambridge, Isaac met another Isaac, Isaac Barrow. Barrow was twelve years Newton's senior and an extremely well-rounded and highly regarded scholar. Prior to 1663, Barrow had a faculty appointment in Greek. However, Reverend Henry Lucas, former Member of Parliament and noted philanthropist, left a number of charitable donations at his death, including four thousand books and a substantial endowment for a faculty position in mathematics at Cambridge. Barrow became the first Lucasian (an endowed chair from the Lucas family) professor of mathematics at Cambridge. His advice to students was to work independently and keep natural philosophy and mathematics at the forefront of their studies. Barrow's private research interests included the mathematics of derivatives (the rate of change of a mathematical function), a subject that Newton absorbed eagerly.

NEWTON'S FIRST CRISIS

Just as Newton was awarded his bachelor's degree, the Great Plague was running rampant through England, so public gatherings were banned and universities were shut down (though church services continued) to minimize the spread of the disease. Newton returned to Woolsthorpe in late 1665. Perhaps because Newton's shiny new degree impressed his mother, she gave up her quest to turn him into a farmer. Newton was free to work on projects of his choice, but this time it wasn't clocks and windmills. Armed with the latest knowledge he could glean from Cambridge's books, and the questions he had swirling around in his head, Newton set himself loftier goals and concentrated on them mightily. "I keep the subject constantly before me," he said, "and wait 'till the first dawnings open slowly, by little and little, into a full and clear light."[3] No one knew how long the university shutdown would last, but it turned out to be almost two years. So, in the intervening time, Newton was able to accomplish quite a lot. He

- invented a mathematical technique called fluxions, which turned out to be equivalent to calculus;
- formulated three general laws explaining the motion of bodies;
- conceived the law of universal gravitation;
- systematized the general procedure by which science operates.

Upon returning to Cambridge, Newton applied to continue his studies by becoming a Fellow of the University (a member of the group at Cambridge with lodging and a stipend), and Isaac Barrow was appointed to examine him to determine his worthiness. It's difficult to imagine Barrow's delight when he saw the fruit of Newton's two years of effort. Newton was admitted to the Fellowship handily and was awarded a master's degree the following year.

NEWTON SCORES

In 1669, Isaac Barrow resigned his faculty appointment to become King Charles II's chaplain. He recommended his professorship post be filled by his prize former student, and so Newton became the second Lucasian Professor of Mathematics. By this time, Newton's major research interest was the nature of light. He had invented and built a telescope using a mirror rather than a lens. Prior telescopes, like the ones used by Galileo, suffered from inaccuracies due to the difficulty in making large lenses. Newton was able to use a curved mirror, which was considerably easier to build and much more accurate. Newton's telescope was superior to any other such instrument in existence at the time. Although Newton was naturally reserved, Isaac Barrow suggested he demonstrate the telescope at a meeting of England's prestigious Royal Society. Until then, Newton had written little and stayed out of the public eye, but Barrow was persuasive and Newton agreed. His telescope caused a sensation at the meeting. Partially in response to the Royal Society's enthusiasm, Newton published his theory about light and color the following year. Things looked rosy for Newton, but they didn't remain that way very long.

NEWTON'S SECOND CRISIS

Almost ten years earlier, the Royal Society (which had originally been the Society for the Promoting of Physico-Mathematical Experimental Learning) had been formed. One of the offi-cers was the curator of experiments, whose job was to demonstrate three or four significant experiments for each (weekly) meeting. At first, the society had no funds to pay a salary for this function or even to buy mate-rials for the experiments. This impos-sible-seeming task was undertaken by Robert Hooke, who became the chief thorn in Newton's side.

Robert Hooke (1635–1705). By Rita Greer, 2004. From Wikimedia Commons, user James.Leek.

Hooke was born on the Isle of Wight, off the southern coast of England, in 1635, the youngest of four children. Since his father and two uncles were ministers, they probably expected Robert to become a cleric also. A sickly child, Hooke was mostly homeschooled and was quite interested in mechanical things, especially clocks. When Hooke was thirteen, his father died, and he used his inheritance to journey to London to become a clockmaker's appren-tice. However, he proved adept at school in London and gave up the tradesman's life to enter Oxford. After studying many different subjects, including learning to play the organ, Hooke began working for the scien-tist Robert Boyle, and helped him formulate Boyle's law of gases. Boyle recommended Hooke to the Royal Society, where he became the curator of experiments.[4]

Hooke was a talented experimenter and inventor who enjoyed the chal-lenge of working on a wide variety of inventions and experiments. One of Hooke's works is well known to physics students today: Hooke's law, which states that an ideal spring's deflection is directly proportional to the force

causing the deflection, and is still used
in physics and engineering. In addition
to his work on watch springs, land surveying, and microscopes, Hooke built
a reflecting telescope (which was far
less effective than Newton's) and had
many ideas of his own about light and
color. When Newton's theories were
published, Hooke was extremely critical of his work. He claimed that the
parts of Newton's theory that were
correct were stolen from him, and that
the other parts were wrong. Newton
didn't take kindly to the criticism and
carried on a semipublic feud with
Hooke that continued for thirty years.[5]

Edmond Halley (1656–1742). By
Thomas Murray (1663–1735). From
Wikimedia Commons, user Hohum.

The Newton/Hooke controversy also extended into another area: gravity.
Although Newton was generally a solitary worker, an ally showed up at a
critical time: Edmond Halley.

FINALLY, SOMEONE TAKES NEWTON'S SIDE

Halley's family had been wealthy soap makers, which meant that they were
able to make sure that Edmond received a first-class education, as well as
lots of astronomical equipment. He excelled in mathematics and astronomy
to the extent that, while still an undergraduate, he became an assistant to the
first Astronomer Royal, John Flamsteed. After helping Flamsteed compile
a star catalogue, Halley was sent to St. Helena (an island in the South
Atlantic, the southernmost British possession at the time) to extend the catalogue to southern hemisphere stars. He was quite successful, and earned
the nickname from Flamsteed "the Southern Tycho." Returning home, he
was elected to membership in the Royal Society and awarded his master's
degree by intercession of King Charles II, since he had completed the star

catalogue rather than mind his studies. At age twenty two, he became one of the society's youngest fellows.[6]

Thanks to a dispute between Robert Hooke and astronomer Johannes Hevelius, Halley was dispatched to Danzig (in modern-day Poland) in order to double-check Hevelius's measurements, about which Hooke suspected inaccuracies. The twenty-four year old Halley vouched for the measuring accuracy of the sixty-eight year old Hevelius, silencing Hooke. Returning home, Halley worked on cometary orbits with Hooke at the encouragement of the great British architect Christopher Wren, but they were unable to analyze the mathematical form of these orbits.

Baroque period gentleman avoiding flying object. Used with permission from Sidney Harris.

Halley then went to see Newton as Flamsteed had done four years earlier to see if he could solve the comet orbit problem. Newton assured him he had worked it all out but had misplaced the details. He promised he would redo it and send the results to Halley when he finished. When Halley got Newton's answer, he realized the significance of the law of universal gravitation: all bodies with mass attract all other bodies with mass, and the force decreases as the square of the distance between bodies increases. He got Newton to expand the work, and Halley himself oversaw its publication as *Principia Mathematica Philosophiae Naturalis*. At the direction of the Royal Society (which had no money), Halley bore the printing costs himself, in spite of his family fortunes being on the wane. The *Principia*, as it became known, included differential and integral calculus (called fluxions by Newton), three general laws of motion, universal gravitation, and more. It established Newton as a preeminent scholar, and generated both controversy and enormous respect.

HOOKE RETALIATES

Hooke was incensed at the publication of the *Principia* and claimed that he knew about gravity before Newton, so he should get the credit.[7] Newton disagreed vehemently, but along came Halley to referee the dispute. Halley worked his magic upon both the combatants, and said in a letter to Newton, ". . . that Mr. Hook [*sic*] has some pretensions upon the invention of the rule of the decrease of Gravity, being reciprocally as the squares of the distances from the Center. He sais [*sic*] you had the notion from him, though he owns the Demonstrations of the Curves generated thereby to be wholly your own."[8] This served to quiet the dispute, but Newton and Hooke never got along well. The final incident in this dispute was that Newton was accused of destroying a portrait of Hooke in the Royal Society offices after Hooke died and Newton became president of the Society.[9]

Bonus Material: Newton/Hooke Internet interview. See To Dig Deeper for details.

Perhaps Halley's reward for his peace-making was his analysis of a short-period comet that appeared in 1682. He predicted it would reappear every 75-76 years. When it was observed next, in 1758, right on schedule, it was promptly named Halley's Comet.

NEWTON AND HALLEY VERSUS FLAMSTEED

Halley himself wasn't immune to controversy. John Flamsteed, who had been appointed Astronomer Royal and Master of the Royal Observatory at Greenwich, had embarked on a long observational project to update and expand the star catalog and planetary tables of Tycho Brahe. Newton wanted Flamsteed's data for the moon because he had left that section of the *Principia* incomplete and was considering updating it. Flamsteed was insistent that the project be finished before the data was released, but Newton was impatient. Finally, Halley obtained the information for Newton and published Newton's gravitational analysis, including some of Flamsteed's data. Flamsteed was furious and was further angered when more of his data were edited by Halley and published by Prince George of Denmark in 1714. Flamsteed managed to burn three hundred of the four hundred copies printed, but his anger toward Halley continued until his death. As one Flamsteed biographer put it, "The last thirty years of Flamsteed's extensive correspondence is infused with vituperative remarks about the man [Halley] who should have been his most natural ally."[10]

NEWTON FACES YET ANOTHER CRISIS

Newton's controversy with Hooke was followed by a longer-distance one with Gottfried Wilhelm von Leibniz.

Gottfried Leibniz's father was a professor of moral philosophy at the University of Leipzig. He died when Gottfried was only six years old but still exerted a strong influence on his son's development by leaving him access to an extensive library. The library was mostly in

Gottfried Leibniz (1646–1716). By Christoph Bernhard Francke (c.1665–1729). From Wikimedia Commons, user Andrejj.

Latin, so young Gottfried became fluent in Latin by age twelve. He entered the university at fifteen and received a bachelor of laws degree four years later. He became a diplomat and wound up in Paris by 1672.

In Paris, Leibniz became friends with physicist Christiaan Huygens, who mentored Leibniz in a self-study of physics and mathematics when he realized how deficient Leibniz's education had been in those areas. In 1673, Leibniz traveled to London and visited the Royal Society. Although there is no record of his meeting Newton, he may have seen some of his early work. The following year, Leibniz developed calculus, using differential notation, and published his work in 1684. Newton had developed calculus using geometrical notation and fluxions in 1666, and he published his work in 1687.[11] (Stay tuned, these dates become important later.)

Leibniz was a prodigious letter writer, whose correspondence numbered tens of thousands of letters addressed to over ten thousand people. Leibniz and Newton corresponded, and when it was finally determined that both mathematical formulations of calculus looked quite different but were equivalent, each acknowledged the work of the other. It appeared that they had developed similar ideas independently. But the question of priority had not arisen—yet. In 1699, Leibniz sent several interesting challenge problems to prominent mathematicians, including Newton. Another person in England did *not* receive those problems and was miffed at being omitted. That person was Nicolas Fatio de Duillier.

DE DUILLIER: FRIEND OR FOE?

Nicolas Fatio de Duillier was born in Switzerland and traveled to Paris in 1682 to study astronomy under

Nicolas Fatio de Duillier (1664–1763). From Wikimedia Commons, user D. H.

Giovanni Domenico Cassini. Later, he traveled to London and became a member of the Royal Society in 1688. He became an extremely close friend of Isaac Newton in 1692. Newton gave him money and offered him a regular allowance if he agreed to stay in Cambridge. Nicolas refused, but he did get a vast sum of money from the Duke of Bedford, whom he tutored.

De Duillier considered himself an equal of Newton, told him of mistakes in the *Principia*, and offered to undertake a second edition. Responding to a perceived insult from Leibniz sent in 1699, de Duillier referred to him as the "second inventor" of calculus. Leibniz, knowing de Duillier was Newton's friend, was incensed. This ignited the controversy between Newton and Leibniz. In 1704, an anonymous review of a paper of Newton's implied that Newton had borrowed the calculus idea from Leibniz. This added fuel to the fire, and when it was discovered that Leibniz had altered the dates on prior papers (he changed 1675 to 1673), that made matters worse. The dispute raged on acrimoniously until 1714 when Leibniz died. The modern view is that the whole matter was a tempest in a teapot. Equivalent ideas about calculus were arrived at independently by two prodigious intellects that had, at best, a tenuous connection.[12]

BEYOND MOTION

After a few years, Newton seemed to lose interest in motion and began working on alchemy and biblical chronology, about which he published many more papers than on scientific topics. Briefly, he became a Member of Parliament and was active in the political life of the university. In 1696, he was appointed to be Warden of the Mint (later Master), moved to London, and worked in the Royal Mint offices located in the Tower of London. Much to the chagrin of established bureaucrats, Newton took the ceremonial position seriously and helped reform the monetary system of England by cleaning out corruption and prosecuting counterfeiters. One reform he instituted involved milling (serrating or engraving) the edges of coins. A favorite trick of thieves was to shave coins to harvest the metal. Coins with milled edges couldn't be shaved without revealing the loss. Newton's personal scientific research had ended, but he was elected head of the Royal Society in 1703, and reelected each year until his death in 1727.

Newton was short of stature and became quite stout in his later years. Fellows of the university were not allowed to marry, and Newton never did. He was always kind to his half-brother and sisters and their children. Newton's niece, Catherine Barton, became his London secretary, which made for a lively and well-run household. Newton referred to 1694 as his "Dark Year," in which he suffered from insomnia and had a serious episode of depression. His absentmindedness was legendary. He was so focused that if he left a room to get a drink for a friend, he might forget to return, and the friend would find him working on some problem. One famous story concerned dismounting his horse at the foot of a steep hill and then arriving at the top of the hill holding the bridle but no horse, which had slipped away unnoticed. Newton became very wealthy and distributed most of his money to family members before his death. He died from a bladder stone, and a subsequent autopsy showed the presence of large amounts of mercury in his body, perhaps a result of his alchemy experiments or some medicinal remedy.

Though Newton was notoriously cantankerous, his assessment of himself in a letter to Hooke is remarkably humble: "If I have seen further it is by standing on the shoulders of giants."[13]

CHAPTER 4

THE BATTLING BERNOULLIS AND BERNOULLI'S PRINCIPLE

All birds need to fly are the right-shaped wings, the right pressure and the right angle.

—Daniel Bernoulli[1]

A ntwerp was a bustling city in the Spanish Netherlands and a cosmopolitan center of world trade. Perhaps Antwerp was becoming too international, for in 1567, King Philip II of Spain sent the Duke of Alba to the Netherlands with a sizable army to reassert the king's authority and compel adherence to the Roman Catholic religion. The duke set up a court called the Council of Troubles. This court condemned over twelve thousand people for either treason or heresy. Many fled, including the Bernoulli family, who were spice merchants. Eventually, they wound up in Basel, Switzerland. Even though their family dynamics were often stormy, several Bernoullis made significant contributions to science and mathematics.

Since many of them have similar first names, the Bernoullis will be referred to by first name and, if necessary, their birth year in parentheses. The intention here is clarity, not inappropriate familiarity.

Nicolaus (1623) was a spice merchant in Basel who inherited the business originally located in Amsterdam. He fathered ten children, including (number 5) Jacob (1654) and (number 10) Johann (1667). Jacob studied philosophy and theology at his parents' insistence but resented it greatly. Although he completed degrees in both, he also studied mathematics and astronomy, which were far more to his liking. This disagreement with authority was the start of a recurring pattern in the family. For the next five years after receiving his theology licentiate in 1676, Jacob studied mathematics and traveled throughout Europe, meeting many of the leading mathematicians. He began correspondences with several of them that lasted many years. His travels took him to England in 1681, where he met Robert Boyle and Robert Hooke. In 1687, Jacob was appointed professor of mathematics at the University of Basel.

A Portion of
The Bernoulli Family Tree

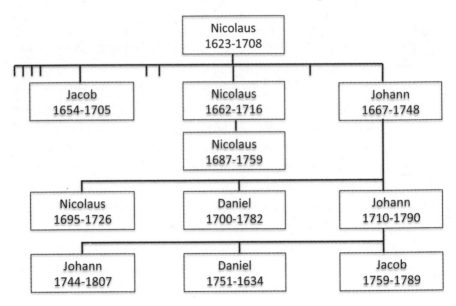

THE SIBLING PLOT THICKENS

Jacob's youngest brother Johann was their parents' last child, so he was directed by them to study business. But Johann protested that business didn't interest him. His father, Nicolaus, was displeased but finally allowed Johann to study medicine as a compromise. Johann entered the University of Basel, where his older brother Jacob taught mathematics. Ostensibly studying medicine, Johann got his brother Jacob to teach him the latest mathematics—calculus—based on the recent (1684) and quite obscure work of Gottfried Leibniz. Johann was a quick study and soon bragged about how he knew more mathematics than his brother, the professor. Jacob wasn't pleased, and the two quarreled, often in public or in print.

After a several-year tour of the Continent, including more than a year tutoring Guillaume Marquis de l'Hôpital in Leibniz's calculus, Johann decided to finish his doctorate and start a family. This meant that he needed a steady job. Since Jacob blocked him from

Johann Bernoulli (1667–1748). By Johann Rudolf Huber (1668–1748) and Johann Jakob Haid (1704–1767). From Wikimedia Commons, user Roybb95~commonswiki.

any faculty position in Basel, he took a teaching job in Groningen, the Netherlands. Johann's first son, Nicolaus (1695), was born several months before he and his wife traveled to Groningen, which turned out to be a difficult journey. Things didn't improve much after they arrived. In France, de L'Hôpital published the first calculus textbook in 1696. Johann received scant mention in the book but swore it was based on notes from his tutorial sessions in France. Since he had been paid (handsomely) for his tutorial work, L'Hôpital believed the notes belonged to him. The book was well received, much to Johann's chagrin.

A NEWTON TRAP?

In 1696, Johann posed a challenge called the brachistochrone problem, which was originally considered by Galileo in 1638: "Given two points

A and B in a vertical plane, what is the curve traced out by a point acted on only by gravity which starts at A and reaches B in the shortest time?"[2]

The problem was published in the scholarly journal *Acta Eruditorum* with a six-month deadline for readers to submit their solutions. Of course, Johann had a solution, so this perhaps seems to have been another way for him to show up his brother in their ongoing feud. But a curious thing happened. No solutions were received by the deadline. Gottfried Leibniz persuaded Johann to extend the deadline to allow foreign competitors to have enough time to work on the problem. Foreign competitors? Who might that be?

Eventually, several solutions were obtained. Jacob Bernoulli solved it (so much for the sibling feud), along with Leibniz and L'Hôpital, plus an anonymous solution published in *Transactions of the Royal Society* in England. Hmmm. Who could that anonymous person be? Perhaps some of the additional wording of the challenge might be relevant here: "There are fewer who are likely to solve our excellent problems, aye, fewer even among the very mathematicians who boast that [they] . . . have wonderfully extended its bounds by means of the golden theorems which (they thought) were known to no one, but which in fact had long previously been published by others."[3]

It appears that Johann Bernoulli (and perhaps Leibniz) was using this problem as a way to test Isaac Newton. After all, Newton had his difficulties with de Duillier and his depression in 1693, and had recently been made Warden of the Mint (recall from chapter 3). Perhaps this was a good time to see if Newton had lost interest in scientific matters and thereby establish Leibniz's priority claim on inventing calculus. Just to make sure Newton saw the challenge, Johann sent two copies of the challenge directly to Newton's home in London in January 1697. According to the later recollection of Newton's niece, here's what happened when the problem arrived: " In the midst of the hurry of the great recoinage, [Newton] did not come home till four (in the afternoon) from the Tower very much tired, but did not sleep till he had solved it, which was by four in the morning."[4]

Newton sent his solution to the president of the Royal Society, where it was published anonymously. Johann wasn't confused by the anonymity of the Royal Society publication. He said, "Tanquam ex ungue leonem" (We recognize the lion by his claw).[5]

Newton also told the Royal Society president, "I do not love to be dunned [pestered] and teased by foreigners about mathematical things."[6]

The calculus feud was reignited seven years later, after Newton had resigned his professorship and devoted himself full time to the Royal Mint.

DIFFICULTIES MULTIPLY

In Groningen, Johann continued his independent streak by introducing experiments into his physics lectures. This managed to offend both Calvinists—who thought sensory data should be used to ascertain God's plan—and Cartesians—who saw sensory information as greatly inferior to reason.

Johann's next child, a daughter, then arrived but lived only a few weeks. This saddened him greatly. Several student disputes of a religious or philosophical nature then swirled around him, in addition to the ongoing controversy with his brother Jacob.

In 1700, Johann's second son, Daniel (1700), arrived and joined this unhappy brood. It wasn't long before the Bernoullis left Groningen. Johann, his wife, their sons Nicolaus and Daniel, and Johann's nephew Nicolaus Bernoulli, who had come to study math with his uncle, all set out for Basel. Johann had finally been offered an academic position—teaching Greek.

During the journey, the family learned that Jacob had died, so when Johann arrived, he lobbied

Daniel Bernoulli (1700–1782).
Used with permission from Sidney Harris.

hard and was rewarded with his brother's teaching position in mathematics. You might think this would be a good time to start playing the "happily ever after" music, but it didn't turn out that way.

After a few years, Johann's third son, also named Johann (1710), arrived. Johann the elder busied himself with his mathematics and spent a lot of time writing letters supporting the Leibniz side of the Newton/ Leibniz controversy (see chapter 3). He was so active that he was referred to as "Leibniz's bulldog."[7]

Soon it became time for Johann to guide his sons' choices for their life's work. The first and favorite son, Nicolaus, was already studying math, and Johann knew there was little financial reward in math. As a result, Daniel was encouraged to study business. At first, he refused but then gave in. Daniel was such a strong student that his father changed his mind and asked him to study medicine. Daniel agreed, but only if his father would teach him mathematics privately. Daniel obtained his PhD in anatomy and botany in 1721. His thesis dealt with the mechanics of breathing, and his work was based partially on his father's ideas about energy conservation. (Does this sound familiar? Genetics strikes again.)

After failing to obtain an academic post at the University of Basel, Daniel made his way to Padua to further his medical training. While traveling through Venice, he became ill and stayed there until he recuperated, working on mathematics and publishing a mathematics book. He also designed an hourglass that would function on ships during heavy seas. For this invention, he won the Paris Academy Prize in 1725. Having acquired a measure of fame, he was offered an academic job. Both Daniel and his older brother Nicolaus were offered and accepted faculty positions in mathematics at a new academy at St. Petersburg. Unfortunately, Nicolaus died within a year after their arrival. Sadness at his brother's death and the harshness of the climate made life difficult for Daniel, but a very interesting thing happened that was based on earlier developments back home. Several years prior, Daniel's father, Johann, did something entirely out of character. His old college roommate, Paul Euler (pronounced *oiler*) had a son, Leonhard, who wanted to study mathematics rather than become a minister like his father. Johann convinced Euler to let his son study mathematics. (Amazing flip-flop?) Just prior to Nicolaus's death, Leonhard Euler finished his doctorate at Basel. When he learned of his son's death, Johann recommended that Leonhard should fill Nicolaus's post at St. Petersburg. Leonhard took the job. Daniel was delighted and offered him a place to live in St. Petersburg. In exchange, Leonhard had to bring brandy, tea, coffee, and other delicacies from home. Thus began an incredibly fruitful collaboration for both of them.

Leonhard Euler's great analytical skills combined with Daniel's superior physical insights produced top-quality work in a wide variety of scientific and purely mathematical areas. These included the harmonic vibrations of a stretched string and the odd harmonics of an open/closed air column, hydrodynamics, and a remarkable anticipation (by one hundred years) of the kinetic theory of gases as the interaction of independent particles.

In spite of the productive collaboration with Leonhard Euler, Daniel was homesick and ready to leave St. Petersburg. In 1733, Daniel and his brother Johann, who had joined him in St. Petersburg, toured the Continent as they journeyed back to Basel. Daniel finally landed an academic position there—teaching botany. (Well, better botany in Basel than sleet in St. Petersburg, right?) Before leaving St. Petersburg, Daniel had submitted an entry to the Paris Academy competition that applied some of his ideas to astronomy. He arrived in Basel to find that he had won the prize. The only problem was that it was awarded jointly to his father, Johann, who had also entered. Johann was furious that his son could be considered his academic equal, so he wouldn't even let Daniel enter the family home.

THE WORST IS YET TO COME

Daniel had left a copy of his major work, *Hydrodynamica*, at the printer's in St. Petersburg, but he continued to polish it for publication. It was eventually issued in 1738. His father also had a book published, *Hydraulica*. Although it wasn't published until 1739, Johann had it predated to 1732. He then claimed Daniel had stolen the work from him. Ironically, the frontispiece of Daniel's work is signed "Daniel Bernoulli, son of Johann."

Johann's plagiarism was noticed before long, and Daniel received appropriate credit for the work. After sixteen years of teaching botany and physiology, Daniel was finally appointed to the chair of physics at St. Petersburg in 1750. He was extremely popular and incorporated many experiments in his lectures. He taught until 1776 and received many honors from eminent scientific societies.

In view of all the contention, jealousy, and dirty tricks of the Bernoullis, one must wonder if the controversies spurred them to greater efforts or whether they would have accomplished more without the distractions. Bernoulli family genes must be quite strong. In 2004, the chairman of the Earth Sciences Department of the University of Basel was Professor Dr. Daniel Bernoulli. In 2015, Facebook lists more than a thousand Bernoullis, and the name "Daniel Bernoulli" has almost seven times as many Facebook "likes" as "Johann Bernoulli."

BONUS EXAMPLES

To illustrate the value of Daniel Bernoulli's work contained in *Hydrodynamica*, let's look at one of his major ideas, the energy conservation principle, and see some examples of it from modern life. Here is Bernoulli's principle of energy conservation from a purely conceptual (nonmathematical and somewhat simplified) viewpoint.

A flowing fluid has three forms of energy: random kinetic energy, represented by pressure; directed kinetic energy, represented by the speed of the flow; and gravitational potential energy, which depends on the height of the fluid. The sum of these three energies is called the total energy, which remains constant under many conditions. Thus, if one form of energy decreases, another must increase, and vice versa. Here are a few examples of this principle in action, as fluid flows or is raised to some height.

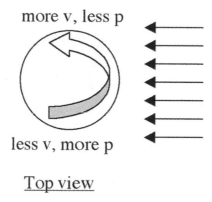

more v, less p

less v, more p

Top view

Curveball

In the diagram, a ball moving from left to right has air flowing across it. If the ball spins counterclockwise as viewed from the top, the ball drags air with it on the left side, increasing its velocity. According to the Bernoulli principle, an increase in velocity must be accompanied by a decrease in pressure, so the pressure on the left side of the ball is smaller. The right side slows the air's velocity, so the pressure on the right side is greater. This imbalance in pressures leads to a net force from right to left, so the ball's path curves to the left. Baseball fans would note that this corresponds to the normal curveball thrown by a right-handed pitcher. If the batter is also right-handed, the ball curves away from him, possibly causing him to swing where the ball isn't. Strike. It's easy enough to figure what would happen for left-handers, and you could check TV slow-motion camera shots for examples. Curveballs also show up in other ball sports such as a soccer, tennis, and Wiffle ball.

Backspin

Look back at the previous diagram and suppose it was a view from the right side of the gallery watching a golfer hit a shot toward the green. The counterclockwise spin causes the golf ball to rise higher than normal. When it lands, the ball would roll forward very little and might even roll backward.

Topspin

In tennis, the idea is often to bring the ball down quickly so that it stays inbounds. To accomplish this goal, the spin direction must be reversed from the previous example, so the path of the ball would dip downward. To apply topspin, tennis players "brush up" the back of the ball as they hit it.

Airplane Wing

In normal flight, an airplane wing's orientation looks like it is going slightly uphill. This is called an *angle of attack*. Since the air must go farther across the top of the wing than the bottom, it must also go faster across the top. From the Bernoulli principle, this means less pressure above the wing and more pressure below. The net force is known as *lift*. Actually, the real situation is much more complicated because the flow is not always smooth (laminar) and turbulence alters things, as does air at the wing ends flowing from the bottom of the wing to the top, creating wingtip vortices.

Sailboat

The sail on a sailboat functions like a wing, and the force generated helps to propel the boat forward. Another component of the force from the sail pushes sideways. Since the sail is high above the boat deck, this other component creates a torque that tends to rotate the boat. To counteract this torque, boats have daggerboards, centerboards, or keels, which generate a torque in the opposite direction. The other possibility for counteracting this torque is called hiking out, as when sailboat passengers or crew hang their bodies out of the boat in the opposite direction to the boat's tilt.

Water Tower

Many communities have water towers, where water is pumped to a high storage location so that it can be distributed later with minimal pumping action. From the standpoint of the Bernoulli principle, this is an example of trading gravitational potential energy for kinetic energy. Pumping the water into the tower often occurs late at night, when there is excess electrical energy available because of minimal demand.

Beach Ball in Blower Stream

You may have seen a beach ball suspended in a vertical air stream from a blower. This demonstration is designed to show the power of the blower, but it also reveals the Bernoulli principle at work. Under these conditions, the ball is remarkably stable. If you try to push it out of the air stream, it is just pushed back toward the center. According to the Bernoulli principle, air within the stream is moving faster than air outside the stream, so the higher pressure outside pushes the ball back.

Chimney

When winter winds howl, smoke from the fire in the fireplace moves up the chimney and out into the air quite smartly. The wind outside causes reduced pressure there, so the higher pressure inside pushes the smoke out.

Here's a handy little experiment you can do to demonstrate the Bernoulli principle at work:

Hold two sheets of paper vertically in front of your face. Blow between the sheets and watch to see if they move toward or away from each other. As the air velocity increases between the sheets, the pressure decreases, so they should move together.

Next time you fly somewhere in an airplane or ride a sailboat, please think kindly about the battling Bernoullis. Their battles won something— for the rest of us.

The Bernoullis generated their own form of heat in terms of family dynamics. The next chapter will deal with heat in the physical realm—different in a sense but still capable of generating hot human interactions.

CHAPTER 5

ANTOINE LAVOISIER AND BENJAMIN THOMPSON (COUNT RUMFORD) HAVE RIVAL THEORIES OF HEAT

If you can't stand the heat, get out of the kitchen.
—Harry S. Truman[1]

Phlogiston? Caloric? You don't hear much about either of these ideas anymore. There is good reason for this. They are abandoned scientific theories about the nature of heat. As we've seen, scientific hypotheses that don't pass the test of experimental evidence must be replaced. As Albert Einstein put it: "No amount of experimentation can ever prove me right; a single experiment can prove me wrong."[2]

Rats and debunked science. Used with permission from Sidney Harris.

Phlogiston and caloric bit the dust through the efforts of some characters you might be interested to meet. But first, we've got to set the scene.

Thanks largely to Galileo and Newton, 1700s physics had escaped the erroneous ideas of the old Greek philosophers and was charting a new course with a solid intellectual framework. But chemistry? Not so much. Chemists were still stuck with the elements fire, air, earth, and water, with some medieval alchemists' arcane rumblings and symbolism tossed in. As an example, combustion and heat were not sorted out as being separate. German scientist Johann Joachim Becher (1635–1682) proposed an idea that was modified by Georg Ernst Stahl (1659–1734), who renamed one of Becher's concepts phlogiston.

Here's how they thought phlogiston fit into the picture: Every combustible substance contains some component of the universal element fire. This component was called phlogiston, from the Greek word for inflammable. For example, if some combustible substance such as wood lost weight when it burned, Stahl figured it was the result of a loss of some of

the wood's phlogiston. Heating also involved phlogiston but was more complicated. We aren't going to expend much energy trying to understand it; the theory is wrong.

Let's meet the fellow who knocked phlogiston for a loop: Antoine-Laurent Lavoisier (1743–1794).

Antoine-Laurent Lavoisier was born in Paris in 1743 to Jean-Antoine and Emilie Punctis Lavoisier. He studied arts, the classics, and science at Mazarin College starting at age eleven. He achieved a law degree at age twenty, following the wishes of his father, but his interest in science prevailed, and he was elected to the Academy of Sciences at age twenty-

Antoine Lavoisier (1743–1794). By Jacques-Louis David (1748–1825) and James Caldwell (1739–1822). From Wikimedia Commons, user Uopchem25NirajPatel.

five. At the same time, he used part of the inheritance from his mother to purchase a share in Ferme générale, a company that advanced estimated taxes to the French government, then collected them from citizens—a practice called tax farming. Lavoisier was an administrator, and one of his colleagues at Ferme générale was Jacques Paulze. In 1771, Paulze's thirteen-year-old daughter received a marriage proposal from Count d'Amerval, who was over fifty and physically repulsive. After threats to his job, Paulze approached Lavoisier with the proposition that he marry Marie-Anne Pierrette Paulze so she could avoid the unwanted advances of Count d'Amerval. Lavoisier had been so studious that his social life was neglected. Besides, Marie-Anne was intelligent, lively, and happened to be very attractive.

Lavoisier accepted. The convent-educated Marie-Anne and Antoine were married in 1771. Several years later, he was appointed gunpowder administrator and built a state-of-the-art chemistry laboratory on the floor above their residence at the Armory. Marie-Anne had become interested in lab work, received formal training, and then assisted Lavoisier in his laboratory research. Additionally, she had artistic skills and drew sketches and engravings of his lab apparatus. Her knowledge of languages was also important, as she translated scientific papers from English to French so her husband could read them. They operated as a well-oiled team.

Marie-Anne Paulze Lavoisier (1758–1836) and Antoine Lavoisier. By Jacques-Louis David (1748–1825). From Wikimedia Commons, user Crisco 1492.

THE END OF PHLOGISTON

The Lavoisiers produced some of the first quantitative chemical measurements, carefully weighing reactants and products in sealed glass vessels. In his 1777 paper *Réflexions sur le phlogistique* (*Reflections on Phlogiston*) (sent to the Royal Academy of Sciences in 1783), Lavoisier attacked the phlogiston theory. His careful measurements showed an increase of mass when substances burned (later understood as oxidation), not the decrease that was predicted by the phlogiston theory. His work was highly regarded and spelled the end for phlogiston. Lavoisier and his wife reportedly burned Stahl's books in celebration. Lavoisier also set the stage for collaboration with noted chemical experimenter Joseph Priestly about what had been called dephloghisticated air (horrid word) as actually being an element, one that Lavoisier named oxygen. He also named another element hydrogen, which combined with oxygen to produce water and which thus could no longer be regarded as one of the primary elements. These findings dealt a serious blow to the old four-element grouping of fire, air, earth, and water, and earned Lavoisier the title of "Father of modern chemistry." Also, by identifying combustion as actually a combination with oxygen, Lavoisier showed that not all heat entails combustion. He called heat a "subtle fluid" that he named caloric. Lavoisier postulated that the total quantity of caloric in the universe is constant and simply flows from hotter bodies to cooler ones. In his experiments, he was one of the first to measure heat flow using an instrument called a calorimeter.

While Lavoisier made significant strides in chemistry, the society around him was in great turmoil. The French Revolution made sweeping changes to French society, including the arrest of all former tax gatherers. Lavoisier was tried, convicted, and guillotined in 1794, along with his father-in-law, Jacques Paulze. Lavoisier's death horrified his scientific colleagues. Mathematician Joseph-Louis Lagrange said, "It took them only an instant to cut off his head, but France may not produce another such head in a century."[3]

BACK IN THE USA

To see the man who had a phenomenal impact not only on Lavoisier's work but also his personal life, we'll have to travel across the Atlantic Ocean. At first, such a connection will seem extremely unlikely, but stay tuned, your patience will be rewarded. That man was Benjamin Thompson (1753–1814).

Benjamin Thompson (1753–1814).
From Wikimedia Commons, user Kelson.

More than a hundred years had elapsed since the Thompson family arrived from England and began farming near Woburn, Massachusetts. They were diligent farmers, well respected and moderately successful. Into this unremarkable family, Benjamin Thompson was born in 1753. Who would have dreamed he would become such a flamboyant character?

Benjamin's father, also named Benjamin, died within two years of his son's birth. Benjamin's mother remarried quickly. Young Benjamin became quite self-reliant at a very early age. Although he excelled at scientific studies and expressed himself with great facility, Benjamin didn't progress very far in school. Yet he resolved he would never be a farmer.

Thompson went to Salem at age thirteen to become a shopkeeper's apprentice. One of his duties involved working with gunpowder, and on one unfortunate occasion a batch of it exploded, seriously injuring Thompson. After he recuperated, he went to Boston to become a doctor's apprentice. That didn't work out either, so he decided to try his hand at teaching. Although he wasn't accepted as a student there, he and his life-long friend Loammi Baldwin sat in on a few scientific lectures at Harvard.

Thompson then moved to Rumford, New Hampshire (now Concord), to take a teaching job. In Rumford, he lived with the new school's headmaster, Timothy Walker. Walker was delighted to have this tall, handsome new teacher with such genteel manners. Sarah Walker Rolfe, the minister's recently widowed daughter, was even more taken with Thompson. She was fourteen years Thompson's senior and, through inheritance from her late husband, the largest landowner in Rumford. Thompson must have also been pleased; within months of his arrival, he was married to Sarah. Moving easily into Sarah's circle of friends, Thompson became almost instantly acquainted with New Hampshire's governor, John Wentworth. The governor was mightily impressed and appointed Thompson to the rank of major in the local militia, the Second Provisional Regiment. Soon, Sarah delivered a daughter, Sarah, whom they called Sally.

THE PLOT THICKENS

With the colonist/loyalist conflict brewing, Thompson's evident sympathy for England and his promotion over older men created great animosity within his militia command. After being charged with "being unfriendly to the cause of liberty," he was acquitted, but his unpopularity grew worse.[4] Finally, he avoided a mob by abandoning his wife and daughter in Rumford. Thompson went to Boston to join the Loyalist forces. He knew the Massachusetts area and the colonial defenses well, so he was invaluable to the British forces. The siege of Boston did not go well for the Crown, and soon the British troops, including Thompson, were evacuated to England. There, Thompson almost immediately linked up with the secretary of state for colonial affairs, Lord Germain. Thompson worked his way up the ladder quickly to become undersecretary of state. With the Revolutionary War raging, doesn't it seem strange that a young former colonial militia officer would be welcomed into the heart of the British war effort, becoming an assistant to the secretary of state? Perhaps he shared his knowledge of the colonists' plans, fortifications, and defenses. In his spare time, Thompson continued to pursue his scientific interests. A particular set of experiments involving improvement of gunpowder secured his election to the Royal Society in 1779. After Lord Germain's forced resignation—he was the least-respected member of prime minister Lord Frederick North's unpopular administration—Thompson was reassigned. Given the rank of lieutenant colonel, he was sent back to the colonies for a brief stint at a British garrison—The King's American Dragoons—in South Carolina and Manhattan. The war concluded soon—in favor of the colonials—and Thompson scurried back to England. For his service to the Crown, Thompson was made a colonel and granted a half-pay lifetime pension.

Since England's war had ended, Thompson decided to seek more military employment. Soon, he departed for Europe where he made contact with Elector Karl Theodor of Bavaria, who was just about to take office in Munich. Amazingly quickly, Thompson became a special confidant to the elector and ultimately became his principal adviser. When he asked for permission from the English military establishment for this new post, it was speedily granted, and Thompson was unexpectedly knighted by King George III. Perhaps the king thought military information from the Continent would be valuable. Of course, information flows both ways, as the elector well knew.

During his time in the elector's service, Sir Benjamin reformed the Bavarian military. Part of his plan was to provide them with decent uniforms and good food. Another part was to rid the Munich streets of beggars by giving them jobs making military uniforms. The elector desired reforms but had been unable to institute any before Thompson arrived, so he was delighted. The lower class was also delighted to get work and food and gave Thompson great support from then on. As another public project, Thompson obtained nine hundred acres of land for a beautiful park on the north end of Munich that became known as English Garden.

In his spare time, Thompson continued his scientific pursuits by performing experiments and writing papers for the *Philosophical Transactions of the Royal Society*. The elector rewarded his efforts by making him a count in 1791. Sir Benjamin chose the title Count Rumford to honor the site of his first success. Before long, political opponents and the threat of war (Austria and France were spoiling for a fight) weighed heavily, so Rumford toured Europe, including England. There, he invented the Rumford Fireplace to please his good friend Lady Palmerston. The new fireplace was vastly more efficient and far less smoky than conventional ones. He also designed improved ovens and stoves.

Back in New Hampshire, Rumford's wife died. Their daughter Sally had reached the age of majority and had just learned of her father's whereabouts. After receiving a letter from Sally, Rumford invited her to visit him. When she arrived, her lack of sophistication displeased him, and he promptly sent her to a finishing school.

A short time later, war raged across the European continent. Austria and France were at war with each other, and both armies were on the move near Munich. Although Munich was technically neutral, the elector fled for his life and sent for Rumford. Rumford returned to Munich with Sally and took charge of the Bavarian Army to organize its defenses. As the Austrians laid siege to Munich in 1796, Rumford rode out of the city to talk to the Austrian commander. Evidently, his well-outfitted and trained army of twelve thousand men convinced the Austrians (and the French) to keep right on marching. He became a local hero, and the elector made Sally Countess Rumford, possibly the first American-born countess.

BORING JOB

In one of his many duties for the elector, Count Rumford supervised a factory where brass cannons were manufactured. Drilling out the barrel was a heat-generating process that required a lot of cooling water. Rumford noticed that the heating continued as long as the drill was turning, even if a blunt drill was used and no metal removed. According to the caloric theory, caloric was leaving the brass, but since the heating effect

Count Rumford. Used with permission from Sidney Harris.

appeared to be limitless, Rumford concluded that there was no such thing as caloric; the mechanical action of the drill had been converted to heat. In 1798, he wrote a paper for the Royal Society, titled "An Experimental Enquiry concerning the Source of the Heat Which Is Excited by Friction." This paper spelled the beginning of the end for Lavoisier's concept of caloric. It required another forty years more for details of the kinetic theory to be worked out by physicist James Prescott Joule, but eventually the kinetic theory fully replaced the caloric hypothesis of Lavoisier. Heat was no longer viewed as a subtle fluid but as simply the motion of molecules.

In 1799, Elector Karl Theodor died. Since he had been roundly disliked by the Bavarians, this appeared to be a good time for Count Rumford to move on. He went back to England and helped found the Royal Institution (not to be confused with the Royal Society, which has similar aims but is older). The first lecturer of the Royal Institution, Humphry Davy (1778–1829) was firmly on the side of the kinetic theory, saying heat was the "vibration of the corpuscles of bodies."[5] Before long, Count Rumford had personality conflicts with the more senior members at the Royal Institution, so he left. He traveled to Paris where he attended entertainments, some of which were given by Lavoisier's wealthy widow Marie-Anne Pierrette Paulze Lavoisier. Their friendship blossomed into marriage, but their styles turned out to be quite different. She enjoyed the social life, while he preferred to tinker with household gadgets.

The quarrels started almost immediately. It was reported that he locked the gates and hid the keys just before one of her parties, and she retaliated by pouring boiling water on his favorite rose collection. Before long, they separated and divorced. Marie-Anne moved out, and Rumford took up residence in the Paris suburb of Auteuil.

Bonus Material: Lavoisier/Thompson Internet interview. See To Dig Deeper for details.

In summary, Benjamin Thompson disproved Antoine Lavoisier's caloric theory, married and divorced his widow, and spent the remainder of his life in Paris. Not bad for a Massachusetts farmer, eh?

FINALLY

So, what was Benjamin Thompson/Count Rumford: soldier, scientist, spy, statesman, Lothario, or what? Biographer Stephen G. Brush has said that Count Rumford "had the most unpleasant personality in the whole of science since Isaac Newton."[6] On the other hand, Rumford donated money for the Rumford Medal to be awarded by the Royal Society (he won the first one); the Rumford Prize, awarded by the American Academy of Arts and Sciences; and the Rumford Professorship at Harvard. Franklin Delano Roosevelt said, in a 1932 interview, "Many-sided men have always attracted me . . . Thomas Jefferson, Benjamin Franklin, and Count Rumford are the three greatest minds America has produced."[7] Roosevelt's admiration was shared by many in England, Bavaria, France, and Massachusetts. But what would Lavoisier have thought about Rumford's demolition of his caloric theory as well as Rumford's marriage and divorce of his beloved Marie-Anne?

Leaving marriage and divorce behind, we will next turn to birth. The periodic table was born, but not without a few pains.

CHAPTER 6

MENDELEEV, MEYER, AND MOSELEY AND THE BIRTH OF THE PERIODIC TABLE

The two most common elements in the universe are hydrogen and stupidity.
—Harlan Ellison, American speculative fiction writer[1]

Chemists ♡ the periodic table. It is a one-stop shop for element symbols, chemical properties, physical properties, atomic mass, atomic number, valences, and plenty of other valuable information. Its one-hundred-plus elements, all arranged in neat rows and columns, might tend to make you suspicious that many clever people were involved in building the table from Aristotle's notion of elements being fire, air, earth, and water. Right. For that matter, how do elements relate to the existence of atoms, denied vehemently by Aristotle (see chapter 1)?

For a long time, scientists ignored this question. Many suspected there were indeed such things as atoms, but they had only vague ideas about what these atoms might look like. Some scientists avoided the question by dealing only with matter in bulk: Whatever particles cannonballs are made of, a cannonball will still travel a distance of x when propelled by an amount of gunpowder y and shot at an angle z. In other words, these scientists acted like physicists. Other scientists focused on interactions between substances: If substance A is mixed with substance B, stirred vigorously and heated, substance C is formed. These scientists started as alchemists, trying to turn base metals into gold, but they later evolved into chemists.

Robert Boyle (Robert Hooke's mentor; see chapter 3), in 1661, defined elements as "certain primitive and simple, or perfectly unmingled bodies;

which are not being made of any other bodies."[2] This was a good start, but it was ignored.

Antoine Lavoisier (see chapter 5) initiated the shift from alchemy to chemistry with his precise measurements using balances. In his book *Traité élémentaire de Chimie* (*The Elementary Treatise on Chemistry*), he listed thirty-three simple substances, twenty-three of which we now recognize as elements. Although his lab work yielded some values of the relative masses, his list of simple substances, which included light and caloric, didn't include masses. His definition of an element was the "last point which analysis is capable of reaching";[3] in other words, the point beyond which no additional parts are discovered. He also named oxygen and hydrogen, but he couldn't bring himself to accept the reality of atoms because he thought they were philosophically impossible.

DALTON TO THE RESCUE

An unlikely fellow showed up around 1800 to put atoms on a more solid footing: John Dalton.

John Dalton was born in 1766 to a modest Quaker family of tradesmen. He was educated in a Quaker elementary school, an alternative to English schools. Because Quakers didn't belong to the Church of England, they were regarded as dissenters. Dalton's older brother had taken over the Quaker school in Manchester, England, known as "New College," and although Dalton was only twelve years old, he assisted his brother by teaching there. He continued teaching until age thirty-

John Dalton (1766–1844). From Wikimedia Commons, user Materialscientist.

four, when the school experienced financial difficulty. He then resigned to become a private tutor. One of his major interests had been meteorology, the study of weather. From his measurements on the properties of air, Dalton supported Democritus's notion of atoms, but he added the idea that atoms differed from each other in size and mass. In a paper read to the Manchester Literary and Philosophical Society in 1803, he said, "An inquiry into the relative weights of the ultimate particles of Bodies is a subject, as far as I know, entirely new." Further, atoms combine with each other in fixed proportions to form compounds in ratios of small, whole

"THE PERIODIC TABLE."

Used with permission from Sidney Harris.

numbers. This is called the law of multiple proportions, which, according to Swiss chemist J. J. Berzelius, would be "a mystery without the atomic theory."

Dalton identified six elements (collections of the same atoms): hydrogen, oxygen, nitrogen, carbon, sulfur, and phosphorus, along with their relative masses, based largely on his and other chemists' measurements of gases.

As we will see, the modern-day list of elements includes more than one hundred different types, none of which turn out to be earth, water, air, fire, or "æther," as Aristotle had claimed (see chapter 1).

Nineteenth-century chemists continued their quest to discover new elements and measure the atomic mass of each element as well as its chemical properties. The lowest atomic mass was defined to be one and was assigned to the element hydrogen (H). (Chemists abbreviate the names of the elements with one or two letters.) In the early 1800s the following elements were known, along with their approximate atomic masses:

lithium (Li), 7	aluminum (Al), 27
beryllium (Be), 9	silicon (Si), 28
boron (B), 11	phosphorus (P), 31
carbon (C), 12	sulfur (S), 32
nitrogen (N), 14	chlorine (Cl), 35
oxygen (O), 16	potassium (K), 39
fluorine (F), 19	calcium (Ca), 40
sodium (Na), 23	titanium (Ti), 48
magnesium (Mg), 24	

LOOKING FOR PATTERNS

First in 1817 and then more thoroughly in 1829, German chemist Johann Döbereiner (1780–1849)—a personal friend of the famous German writer Johann Goethe—published articles in which he examined the properties of sets of elements that he called *triads* (for example, lithium, sodium, and potassium). The elements of each triad have similar chemical properties, and the atomic mass of the second element of the triad is approximately equal to the average of the atomic masses of the other two elements. For example, lithium 7 + potassium 39 = 46. Half of 46 is 23, the mass of sodium. The same relationship worked for two other groups of elements. Döbereiner thought this too strange to be a mere coincidence. He felt that there must be some kind of pattern involved. He proposed that all elements could be grouped in such triads, however subsequent attempts to expand the concept to other groupings were unsuccessful. His attempt failed to provide a comprehensive framework, in part because the masses of elements were measured inaccurately and in nonstandard ways.

Another attempt to organize the elements into a pattern came from an unlikely source: French geologist Alexandre-Émile Béguyer de Chancourtois. In 1862, he arranged the known elements on a cylinder in order of increasing atomic mass and noted that sometimes similar elements having similar chemical properties recurred periodically. Unfortunately, the paper describing his system left out the diagram and included many geological ideas, and so was ignored by chemists of the time.

In the years 1863 to 1866, English chemist John Newlands (1837–1898) proposed another idea: the law of octaves. Newlands stated that when the elements are listed in increasing atomic mass, the eighth element is similar in chemical properties to the first, the ninth to the second, and so on, just like notes in musical octaves. Unfortunately, Newlands's "law of octaves" did not seem to work for elements heavier than calcium, and so his idea was publicly ridiculed. At one scientific meeting, Newlands was asked why he didn't just arrange the elements in alphabetical order instead of by atomic mass, since that would make as much sense as his octaves. Newlands had carried the idea of the metaphor too far; the actual relationship is not so simple. His work was therefore not taken seriously by other chemists.

WILLIAM ODLING'S OFT-FORGOTTEN CONTRIBUTION

In the 1860s, a highly respected academic English chemist, William Odling (1829–1921), devised a table of elements using repeating units of seven elements before starting a new row. Some elements were omitted, however, without any reasonable explanation. Odling considered the scheme to be a simple convenience, and he was not committed to atomism, so the masses of elements didn't play a large role in his scheme. He did not receive recognition on par with other researchers because he didn't advocate for his system strongly enough, so his colleagues regarded it as mere speculation.

THE FIRST INTERNATIONAL CHEMISTRY CONFERENCE

The Karlsruhe Congress of September 1860 brought European chemists together for the first time to discuss many important matters of a chemical nature, including the standardization of atomic masses (also referred to as atomic weights). This meeting, which may have sparked the eventual organization of the International Union of Pure and Applied Chemistry (IUPAC), inspired many to attempt to deal with finding organizational patterns in the ever-increasing number of chemical elements. Among those in attendance were two who had a profound effect on finding those patterns: Julius Lothar Meyer and Dmitri Mendeleev.

Julius Lothar Meyer (1830–1895).
From Wikimedia Commons, user Kelson.

Julius Lothar Meyer was born in Varel, Germany, in 1830 to an academically oriented family. His father was a doctor, and his mother's family also included physicians. He studied at Zurich, Würzburg, and

Heidelberg, earning a medical degree and a PhD in chemistry along the way. His dissertation dealt with carbon monoxide in the blood, and he noted the combination of oxygen with hemoglobin in the blood.

Soon after attending the Karlsruhe Congress in 1860, Meyer wrote a book, *Die modernen Theorien der Chemie* (*The Modern Theory of Chemistry*). At 164 pages, it is more like a long treatise than a textbook, but this 1864 work includes a table of elements that exhibit a periodic repetition of chemical properties as the atomic mass increases. Befitting the slim nature of the publication, the table is short: it contains only twenty-eight elements. Meyer was a proponent of atomism, but he clearly chose not to make any predictions about gaps in the table representing missing elements. As time moved along, Meyer augmented and revised this book and used it as a teaching tool, mostly for graduate students. The next time the table was revised was in 1869, but its publication didn't occur until 1870. The updated periodic table was almost identical to another table that had been published in 1869. So, where did the other table come from; who beat Meyer into print by a few months?

MENDELEEV
TO THE RESCUE

Of all the six "periodic" tables introduced in the period between 1860 and 1870 the last one came from the Russian chemist, Dmitri Mendeleev (1834–1907). Dmitri Mendeleev was the last child (of eleven to sixteen; the number is unclear) born to Ivan Pavlovich Mendeleev and Maria Dmitrievna Mendeleeva (née Kornilieva) near Tobolsk in Siberia. His father taught fine art, politics, and philos-

Dmitri Mendeleev (1834–1907). From Wikimedia Commons, user Materialscientist.

ophy at the local gymnasium, but he developed cataracts and lost his sight and his job. His mother restarted an abandoned glass factory owned by the family and operated it for several years until it burned down. Because Mendeleev demonstrated early scientific talent, his mother set out to take him to Moscow, the home of her wealthy brother, so he could continue his education at a first-class institution. The social turmoil in Moscow (revolts in nearby Romania and Hungary) prevented non-Muscovites from being admitted to the university. Mendeleev, his sister, and his mother continued their search for education in the Russian capital, St. Petersburg, but met the same fate there. Finally, at the Main Pedagogical Institute in St. Petersburg, a professor who knew Mendeleev's father when he attended there recommended Mendeleev, and he was admitted. Mendeleev endured a rocky start as a student at the Institute. His mother and sister contracted and died of tuberculosis, and he, too, became ill. In spite of poor early grades; his illness, which necessitated classmates bringing him books in the hospital; and his uncontrollable temper; Mendeleev finished at the top of his class.

In one of his later works, he made this dedication: "This investigation is dedicated to the memory of a mother by her youngest offspring. Conducting a factory she could educate him only by her own work. She instructed by example, corrected with love, and in order to devote him to science she left Siberia with him, spending thus her last resources and strength."

This education, however hard-won, prepared him to teach at the gymnasium level as his father did, a prospect he did not relish. After a brief teaching stint in Odessa (in Ukraine), Mendeleev applied to the Main Pedagogical Institute to work on his magister degree (similar to a master's degree), but the Institute had shut down, so he applied to the St. Petersburg Imperial University. He was accepted and was awarded his first and second Magister degrees in chemistry within a year. For the next two years, Mendeleev gave seminar lectures and supervised undergraduate laboratory studies, all at an inadequate salary. His work was sufficiently impressive that the university granted him a twenty-two-month fellowship to study abroad. After touring Europe for a while and meeting many influential chemists, Mendeleev settled on Heidelberg as a place to study. There he met Robert Bunsen (1811–1899), but Bunsen's lab lacked the equipment for the research Mendeleev was doing, so he bought the needed equipment and set up a lab in his apartment. In Heidelberg, Mendeleev met fellow chemists Emil Erlenmeyer (1825–1909) and Alexander Borodin (1833–1887). In addition to his career in chemistry and medicine, Borodin became famous as a symphonic and operatic composer.

Mendeleev met many other chemists at Karlsruhe Congress, possibly including Julius Lothar Meyer, but there is no specific record of their meeting. Denied any extension of his education abroad, Mendeleev returned to St. Petersburg, just as the serfs were becoming emancipated. Although he got an adjunct teaching position, societal and student unrest closed the university. He needed to do something to avoid starving, so he wrote a textbook on organic chemistry. The book was five hundred pages long and was completed within sixty-one days. The hurry was occasioned by the deadline to compete for the Demidov Prize (the Russian equivalent of the Nobel Prize) from the Academy of Sciences. The book was completed in time, and Mendeleev won the prize in 1862. The prize money enabled him to marry Feozva Nikitichna Lescheva. Although strongly encouraged by Mendeleev's sister Olga, it was an extremely unhappy union in which the two eventually were unable to tolerate the presence of each other in the same house.

Mendeleev's career prospered, however. He completed his PhD dissertation and taught at several institutions in St. Petersburg, winding up at St. Petersburg State University. With all the societal reforms, higher education became much more popular, and Mendeleev's freshman chemistry course became quite crowded. Further, there were no suitable texts, so Mendeleev set out to write a Russian-language text for inorganic chemistry. The six hundred pages of volume 1 dealt with only eight elements: hydrogen, oxygen, carbon, nitrogen, fluorine, chlorine, bromine, and iodine. Since he had the entire rest of the elements to deal with in volume 2, Mendeleev knew he had some serious organizing to do. He spent about six weeks at the start of 1869 experimenting with various arrangements of the remaining elements.

I began to look about and write down the elements with their atomic weights and typical properties, analogous elements and like atomic weights on separate cards, and this soon convinced me that the properties of elements are in periodic dependence upon their atomic weights.[4]

The lightest element, hydrogen, was the sole member of the first row. The second row began with lithium and continued all the way to fluorine. When Mendeleev reached the next element, sodium, an element whose properties were quite similar to those of lithium (e.g., reacting violently with water), he started a new row, placing sodium below lithium. He continued along this row until the pattern repeated with potassium (which also reacted violently with water), which he placed below sodium. Next, he placed calcium below magnesium and beryllium.

H 1 Hydrogen						
Li 7 Lithium	Be 9 Beryllium	B 11 Boron	C 12 Carbon	N 14 Nitrogen	O 16 Oxygen	F 19 Fluorine
Na 23 Sodium	Mg 24 Magnesium	Al 27 Aluminum	Si 28 Silicon	P 31 Phosphorus	S 32 Sulfur	Cl 35 Chlorine
K 39 Potassium	Ca 40 Calcium					

MIND THE GAP

After calcium, the next known element in order of atomic mass was titanium. If Mendeleev placed titanium immediately following calcium, it would occupy a place directly below aluminum. But Mendeleev knew from his study of the chemical properties of boron and aluminum that titanium did not fit into that group: boron and aluminum form compounds with oxygen in which the ratio of boron or aluminum atoms to oxygen atoms is two to three. Using subscripts to indicate the relative number of atoms, these compounds can be represented by the formula E_2O_3, where E represents an element in the boron-aluminum group. Titanium forms a compound with oxygen, whose general formula corresponds to that of the carbon-silicon group, namely, EO_2. Because titanium's chemical properties more closely matched those of the elements in the carbon-silicon group than those in the boron-aluminum group, Mendeleev boldly skipped one space, placing titanium below silicon, as shown below:

H 1 Hydrogen						
Li 7 Lithium	Be 9 Beryllium	B 11 Boron	C 12 Carbon	N 14 Nitrogen	O 16 Oxygen	F 19 Fluorine
Na 23 Sodium	Mg 24 Magnesium	Al 27 Aluminum	Si 28 Silicon	P 31 Phosphorus	S 32 Sulfur	Cl 35 Chlorine
K 39 Potassium	Ca 40 Calcium		Ti 48 Titanium			

This space or gap was actually a prediction. Mendeleev's hypothesis was that the chemical properties of the elements recur in a periodic fashion or, more specifically, that the chemical properties of the elements are periodic functions of their atomic masses. Based on this hypothesis, Mendeleev predicted that another element should exist that fits the blank space in the periodic array of elements. This element should have properties similar to boron and aluminum and should have atomic mass between those of calcium (40) and titanium (48). His prediction was found to be correct when scandium (Sc), atomic mass 45, was discovered in 1879.

H 1 Hydrogen						
Li 7 Lithium	Be 9 Beryllium	B 11 Boron	C 12 Carbon	N 14 Nitrogen	O 16 Oxygen	F 19 Fluorine
Na 23 Sodium	Mg 24 Magnesium	Al 27 Aluminum	Si 28 Silicon	P 31 Phosphorus	S 32 Sulfur	Cl 35 Chlorine
K 39 Potassium	Ca 40 Calcium	Sc 45 Scandium	Ti 48 Titanium			

Mendeleev also predicted the existence of two other as yet undiscovered elements, which he called ekasilicon and ekaluminum. In this sense, Mendeleev's periodic table is a hypothesis, and the skipped spaces are predictions of missing elements. As we have seen, experimental evidence that matches predictions lends support to the hypothesis. Besides the discovery of scandium in 1879, the other two elements predicted, called germanium and gallium, were found in 1886 and 1875, respectively.

REENTER MEYER

Meanwhile, in Germany, Julius Lothar Meyer had been working on revisions to his table of elements, and several versions predated Mendeleev's. So, why didn't Meyer get credit? Two reasons: First, Mendeleev's publication of the complete table preceded Meyer's, a fact alluded to in Meyer's book. Second, Mendeleev made his predictions with much more specificity about the chemical and physical properties of the predicted elements. Although a few of Meyer's German colleagues pressed the case for his priority, he himself never did. One interesting event from 1882 was that Mendeleev and Meyer shared the British Royal Society's Davy Medal in

1882 "for their discovery of the periodic relations of the atomic weights."
Julius Lothar Meyer became the first chemistry professor at the University
of Tübingen and taught there until his death in 1895.

SPECIAL DISPENSATION

Mendeleev wasn't quite done. In 1876, while still married to Feozva, he
met seventeen-year-old Anna Ivanova Popova, his niece's best friend.
This beautiful young woman became an obsession with Mendeleev, and
he resolved to marry her or else jump into the ocean and drown. Unable to
receive an immediate divorce through the Orthodox Church, Mendeleev
nevertheless found a priest to marry him and Anna. He avoided prosecution
as a bigamist by appealing to the czar. When another nobleman desirous of
the same dispensation also appealed to Czar Alexander, alluding to Mendeleev, the czar is reported to have replied, "Mendeleev has two wives,
yes, but I have only one Mendeleev."[5]

In later years, Mendeleev's eccentric tendencies became even more
evident. His piercing stare riveted his (mostly undergraduate) students, and
his beard and long hair (trimmed yearly) added to his image. Mendeleev's
strong democratic beliefs also flourished. He delivered a student protest
to the Ministry of Education and was rewarded with an official rebuke.
In 1890, he resigned his teaching post and entered government service.
Mendeleev worked for the modernization of Russia in many ways, rising
to the position of director of the Central Board of Weights and Measures.
Does this remind you of Isaac Newton (chapter 3) who became Master of
the Royal Mint?

NOBEL PRIZE CONTROVERSY

In 1905 Mendeleev was elected a member of the Royal Swedish Academy
of Sciences. The following year the Nobel Committee for Chemistry recommended that the Swedish Academy award the Nobel Prize in Chemistry for 1906 to Mendeleev for his discovery of the periodic table. The

Chemistry Section of the Swedish Academy supported this recommendation. The academy was then supposed to approve the committee choice as it had done in almost every case.

At the full meeting of the academy, a dissenting member of the Nobel Committee, Peter Klason, proposed an alternate candidate for the prize: Henri Moissan, who had isolated the element fluorine. Another influential academy member, Svante Arrhenius, argued that the periodic table concept was too old to win the prize in 1906. Some suggested Arrhenius still held a grudge against Mendeleev for his criticism of Arrhenius's dissociation theory, which said that acids dissolve in water to form two charged substances called ions. Remarkably, Moissan's nomination won the majority vote of the academy.

In 1907, Mendeleev died from influenza at the age of seventy-two in Saint Petersburg. The large impact crater Mendeleev on the far side of the moon, as well as element number 101, the radioactive mendelevium, are named after him. When he died, students carried the periodic table in the funeral procession.

MOSELEY'S ATOMIC NUMBERS TO THE RESCUE

Despite its usefulness, Mendeleev's periodic table was based entirely on empirical observation supported by very little understanding. It turns out that a measure unknown in Mendeleev's time, the *atomic number*—the number of protons in the nucleus of an atom—is a more fundamental guide to correlating chemical properties than atomic mass.

The person who contributed

Henry Moseley (1887–1915).
From Wikipedia, user Deglr6328.

this vital piece of information was a young British physicist named Henry Gwyn Jeffreys Moseley (1887–1915).

H. G. J. Moseley was born in the town of Weymouth, England, in 1887. He was educated in private schools and won a scholarship to Eton College, probably Britain's most prestigious prep school. At age eighteen, he won Eton's physics and chemistry prize. In 1906, he was admitted to the University of Oxford's Trinity College, where he studied physics. Then, in 1910, he moved to the University of Manchester to join Ernest Rutherford's research group.

In 1913, Moseley celebrated his twenty-sixth birthday. Mendeleev's periodic table was forty-four years old and had grown in size as new chemical elements were discovered and added to it. However, a basic flaw remained in Mendeleev's table: the position predicted by an element's atomic mass did not always match the position predicted by its chemical properties.

Moseley began to examine the question of whether elements could have a more fundamental property than atomic mass. He had learned from William and Lawrence Bragg (unique father-and-son Nobel prize–winners for X-ray studies that led to crystallographic studies that determined DNA structure; see chapter 13) that when high-energy electrons hit solids such as metals, the solids emit X-rays. He wondered if he could study these X-rays to learn more about what goes on inside atoms.

Moseley put together an experimental apparatus to shoot high-energy electrons at different chemical elements and measure the wavelength and frequencies of the resulting X-rays. He discovered that each element emits X-rays at a unique frequency. His data made most sense if the positive charge in the atomic nucleus increased by exactly one unit as each element in Mendeleev's periodic table is examined in its turn. This meant that Moseley had discovered that the basic difference between elements is the number of protons they contain.

He said, "We have here a proof that there is in the atom a fundamental quantity, which increases by regular steps as one passes from one element to the next. This quantity can only be the charge on the central positive nucleus, of the existence of which we already have definitive proof."[6]

His hypothesis about the placement of each element in a periodic table was that it corresponded to its atomic number. Argon, for example, although having an atomic mass greater than that of potassium (39.9 versus 39.1, respectively), was placed *before* potassium in the periodic table. While analyzing the frequencies of the emitted X-rays, Moseley noticed that the atomic number of argon is 18, whereas that of potassium is 19, which indicated that they were indeed placed correctly. Moseley also noticed three gaps in his table of X-ray frequencies, so he predicted the existence of three unknown elements: rhenium, discovered in 1925; technetium, discovered in 1937; and promethium, discovered in 1945.

In its modern form, the periodic law can be stated thusly: The chemical properties of the elements are periodic functions of their atomic numbers. Tellurium, atomic number 52, thus precedes iodine, atomic number 53. In a sense, Mendeleev was lucky, for increasing atomic mass is almost always correlated with increasing atomic number.

1 H 1 Hydrogen							
3 Li 7 Lithium	4 Be 9 Beryllium	5 B 11 Boron	6 C 12 Carbon	7 N 14 Nitrogen	8 O 16 Oxygen	9 F 19 Fluorine	
11 Na 23 Sodium	12 Mg 24 Magnesium	13 Al 27 Aluminum	14 Si 28 Silicon	15 P 31 Phosphorus	16 S 32 Sulfur	17 Cl 35 Chlorine	
19 K 39 Potassium	20 Ca 40 Calcium	21 Sc 45 Scandium	22 Ti 48 Titanium		34 Se 79 Selenium	35 Br 80 Bromine	
	Note: Atomic numbers are shown above element symbols; atomic masses below.				52 Te 128 Tellurium	53 I 127 Iodine	

When World War I began in 1914, Moseley enlisted as a volunteer in the British Army's Royal Engineers. His family pleaded with him to continue his scientific research, and the army was reluctant to accept him. Moseley had to fight hard to get into the army.

Second Lieutenant Henry Moseley was killed in battle by a sniper at the age of twenty-seven in Gallipoli, Turkey, on August 10, 1915. His grave is located on Turkey's Gallipoli Peninsula. As a result of Moseley's death, and after much lobbying by Ernest Rutherford, the British government placed a ban on other scientists of repute serving in wartime front-line roles.

Robert Millikan (1868–1953), winner of the Nobel Prize in Physics in 1923, wrote: "Had the European war had no other result than the snuffling out of this young life, that alone would make it one of the most hideous and most irreparable crimes in history."[7] In 1916 no Nobel Prizes were awarded in physics or chemistry. There is a strong consensus that Henry Moseley, had he been alive, would have received one of these awards. (Nobel Prize–winners must be alive at the time of presentation.)

ROOM TO GROW

The number of known elements has increased greatly since the days of the ancient Greeks. There are now 118 known elements, each of which is described quite well by the periodic law. In order to accommodate all of these elements, the periodic table has been modified in form and expanded. As a result, scandium is no longer placed under boron and aluminum, and titanium is no longer located under carbon and silicon.

1	2	3	4	5	6	7	8	9	10	11	12	13	14	15	16	17	18
H (1)																	He (2)
Li (3)	Be (4)											B (5)	C (6)	N (7)	O (8)	F (9)	Ne (10)
Na (11)	Mg (12)											Al (13)	Si (14)	P (15)	S (16)	Cl (17)	Ar (18)
K (19)	Ca (20)	Sc (21)	Ti (22)	V (23)	Cr (24)	Mn (25)	Fe (26)	Co (27)	Ni (28)	Cu (29)	Zn (30)	Ga (31)	Ge (32)	As (33)	Se (34)	Br (35)	Kr (36)
Rb (37)	Sr (38)	Y (39)	Zr (40)	Nb (41)	Mo (42)	Tc (43)	Ru (44)	Rh (45)	Pd (46)	Ag (47)	Cd (48)	In (49)	Sn (50)	Sb (51)	Te (52)	I (53)	Xe (54)
Cs (55)	Ba (56)	57–70 Lu (71)	Hf (72)	Ta (73)	W (74)	Re (75)	Os (76)	Ir (77)	Pt (78)	Au (79)	Hg (80)	Tl (81)	Pb (82)	Bi (83)	Po (84)	At (85)	Rn (86)
Fr (87)	Ra (88)	89–102 Lr (103)	Rf (104)	Db (105)	Sg (106)	Bh (107)	Hs (108)	Mt (109)	Uun (110)	Uuu (111)	Uub (112)		Uuq (114)				

*Lanthanide series

La (57)	Ce (58)	Pr (59)	Nd (60)	Pm (61)	Sm (62)	Eu (63)	Gd (64)	Tb (65)	Dy (66)	Ho (67)	Er (68)	Tm (69)	Yb (70)

**Actinide series

Ac (89)	Th (90)	Pa (91)	U (92)	Np (93)	Pu (94)	Am (95)	Cm (96)	Bk (97)	Cf (98)	Es (99)	Fm (100)	Md (101)	No (102)

Modern periodic table. Courtesy of wpclipart.com.

THE NAME GAME

In the past, the group that first synthesized an element got the privilege of naming it. Recently, however, rules regarding the nomenclature of new elements have been enacted to prevent acrimonious conflict over who gets to name a newly synthesized element.

A case in point deals with the naming of element 105, the element containing 105 protons in its nucleus. The Soviets may have synthesized a few atoms of element 105 in 1967 at the Joint Institute for Nuclear Research in Dubna, Russia, USSR, but because the Dubna group did not propose a name for the element at the time they announced their preliminary data—a practice that has been customary following the discovery of a new element—it was surmised by American scientists that the Soviets did not have strong experimental evidence to substantiate their claims. Soviet scientists contended, however, that they did not propose a name in 1967 because they preferred to accumulate more data about the chemical and physical properties of the element before doing so. After completing further experiments, they proposed the name nielsbohrium.

In 1970, a group of investigators at the Lawrence Radiation Laboratory of the University of California at Berkeley announced that they could
not duplicate the Soviet experiment but were able to produce an isotope
(another form of the same element with a different number of neutrons
in the nucleus) of element 105. The Americans referred to this isotope as
hahnium-260 because they wanted the new element to be named in honor
of Otto Hahn, the discoverer of nuclear fission (see chapter 12). (The 260
refers to the number of protons and neutrons in the nucleus of this isotope
of element 105.)

In 1985, the International Union of Pure and Applied Chemistry
(IUPAC) decided to set up an ad hoc working group to consider the competing claims for priority of discovery of elements 101 through 112. The
group first met in Bayeux, France, in February 1988. It published its final
report five years later in August 1993.

At its thirty-eighth general assembly, held in 1995 at the University of
Surrey in Guildford, England, the IUPAC decided to reconsider its recommended names. Following a further two years of consultation, the union
ratified a slate of names for elements 101 through 109 at its thirty-ninth

*"What do you expect, since 90% of all the
scientists who ever lived are alive today?"*

Used with permission from Sidney Harris.

general assembly in Geneva in 1997. The names met with widespread approval. Element 105 was dubbed dubnium (symbol Db) after the name of the city where the Soviet scientists worked.

Where are all these conflicts coming from?

As if element conflicts aren't enough, the next chapter will deal with electrical conflicts. The excitement continues.

CHAPTER 7

Used with permission from Fugene Mann.

WESTINGHOUSE AND TESLA VERSUS EDISON— AC/DC TITANS CLASH

If Edison had a needle to find in a haystack, he would proceed at once with the diligence of a bee to examine straw after straw until he found the object of his search . . . I was a sorry witness of such doings, knowing that a little theory and calculation would have saved him ninety percent of his labor.

—Nikola Tesla[1]

Genius is one percent inspiration and ninety-nine percent perspiration.

—Thomas Alva Edison[2]

Tesla has contributed more to electrical science than any man of his time.

—Lord Kelvin[3]

A merica in the 1880s resembled a giant, recently awakened and just beginning to flex some muscle. There were big-time entre-preneurs in railroads, banking, steel, coal, oil, and many other developing industries, and there were constant struggles for their control. The conflict we're about to discuss had a lot of industrial overtones, but its roots were in the scientific world.

Let's start with a look at some of the participants, each of whom came from humble origins but whose egos had grown to be huge by the time the clashes took place.

PARTICIPANT 1: THOMAS ALVA EDISON

Thomas Alva Edison (1847–1931) was born in Milan, Ohio, and was the seventh and last child of Samuel Edison and Nancy Matthews Elliott Edison. In 1854, the Edisons moved to Port Huron, Michigan, where they hoped the lumber business would be better than it was in Ohio. Shortly after the move, young Alva (as he was called) contracted scarlet fever, which delayed his entry into formal schooling and possibly left him with hearing loss. At age eight, when he finally got to school, he came home crying after three months, saying his teacher had referred to him as "addled." His mother with-

Thomas Alva Edison (1847–1931).
Used with permission from Sidney Harris.

drew him immediately and took on the task of his education. She encouraged his independent thinking, which flowered under her guidance. Edison worked his way through many classics, including Newton's *Principia*, with the help of a family friend. He acquired a distaste for mathematics, saying he thought Newton could have appealed to a wider audience if he used less math. He loved to tinker.

In 1859, the Grand Trunk Railroad completed a rail line that included a run from Port Huron to Detroit. Edison worked on the train, selling Port Huron farm produce while on the way to Detroit, and Detroit newspapers on the way back. Edison's railroad career was quite successful. He was even allowed to use a spare freight car as a laboratory for his personal experiments. He also printed a newspaper of his own design, the *Grand Trunk Herald*. A chemical fire in the railway car laboratory, however, got Edison kicked off the train. While still working for the railroad, Edison saved three-year-old Jimmie MacKenzie from being hit by a runaway boxcar. Jimmie's grateful father, a station agent, taught Edison to operate the telegraph machine, an extremely useful skill.

By 1863, Edison had become an itinerant telegraph operator. Moving from place to place to substitute for operators who had gone to war, Edison worked as an operator for several years. He usually volunteered for the night shift so he could study and tinker during the day.

A visit home in 1868 revealed his family enduring hard times, so Edison realized he needed to take better control of his life. He moved to Boston and began working for Western Union as a telegraph operator. By the end of the year, he made a fateful decision. He resigned from Western Union to devote his full effort to "bringing out inventions."

His first invention was a vote recorder for the Massachusetts legislature. However, legislators were uninterested in rapid vote recording. They actually wanted less speed so they could have more time to discuss, filibuster, and persuade their colleagues. Edison's first invention was thus abandoned as a failure.

His next effort, an improved stock ticker, was only a little better. A system for sending more than one message over a single telegraph wire (called a *duplex*) seemed promising, so Edison borrowed $800 to build the equipment. He then convinced the Atlantic & Pacific Telegraph Company to let him give it a test over their wires, sending multiple messages from Rochester to New York. It failed miserably, even though the next issue of the National Telegraphic Union's magazine, the *Telegrapher*, reported it as a "complete success."

Edison left Boston for New York and arrived there flat broke and in need of a job. In New York, he contacted his friend, Franklin Pope, a well-respected telegrapher, author, and editor of the *Telegrapher* (and possibly a participant in the duplex test). Pope was also chief engineer for the Laws Gold Reporting Company, which ran a service that relayed gold prices via wire from the New York Gold Exchange to several hundred brokers' offices. The company had no job for Edison, but Pope arranged it so Edison could sleep in the company's basement battery room until he found work. Having no other place to go, Edison accepted. After business hours, he had the run of the place and soon figured out how the machinery worked. In the middle of a business day, the transmitter quit, and the office became crowded with messengers from the brokerage houses, wanting current gold prices. Although he wasn't an employee, Edison was present and almost immediately found and repaired the problem (a broken spring). The following day, Edison was made Pope's assistant by the company's owner, Dr. Samuel Laws. Within a month he had Pope's job (at $300 per month) when Pope resigned to become an independent consultant. Within three months, Edison improved the operation and applied for some patents. He found himself an employee of Western Union again when that company bought the Laws Gold Reporting Company in 1871.

Through several more inventions and improvements on telegraphic and printing equipment, including waxed paper for mimeograph machines, Edison was rewarded with a huge payoff from Western Union: $40,000 (almost $8 million in 2015 dollars). With this money, Edison set up his first workshop in Newark, New Jersey. More inventions followed, mostly dealing with telegraphic equipment. Edison also took time to marry and start a family. Their first child, Marion, was born in 1872, followed by a son, Thomas Jr. in 1876. They were nicknamed Dot and Dash. Eventually, Edison and his assistants outgrew the early lab spaces and moved to Menlo Park, New Jersey. Within a year of the move, Edison invented the phonograph, and a year later he was hard at work on an *incandescent* lamp. The principle was simple enough: run an electric current through a material that would heat up and glow brightly enough to provide light. The difficulties included finding a material for the filament that would last, attaching the electrical contacts to the filament securely, finding the right shape for the bulb, and maintaining a vacuum inside the bulb to prevent the hot filament from reacting chemically. Edison's assistants tested a huge number of materials before settling on carbonized thread in a highly evacuated bulb. By late December 1879, a hastily rigged system of electric power generation, distribution, and lighting was set up for public viewing at Menlo Park. Since the current flowed only in one direction, it was called direct current, or DC. The system was a huge success, and people braved stormy weather to see "The Wizard of Menlo Park's" latest invention.

Converting the preliminary system to a commercially viable one took a bit of doing, as well as some time. A dynamo had to be set up to generate the current, bulbs had to be manufactured, and shallow tunnels needed to be dug for the wiring to be buried. One major difficulty was that current flowing through the wires heated them up, so energy was lost in transmission. As long as the customers were located within a mile of the generator, the losses were small. In September of 1882, Edison's Pearl Street Station in Manhattan was finally ready. Before the end of the year, 2,400 Edison bulbs glowed brightly in offices within New York's financial district. But everything wouldn't remain so rosy. Another participant was about to enter Edison's life.

PARTICIPANT 2: TESLA

In 1856, Nikola Tesla was born in Croatia in eastern Europe. His father, Milutin, was a Serbian Orthodox priest, and his mother, Đuka Mandić Tesla, was the daughter of another Serbian Orthodox priest. When Nikola was five years old, he found his older brother's dog dead by the roadside. Dane, his twelve-year-old brother, had been recognized as a child prodigy

Nikola Tesla (1856–1943). Used with permission from Sidney Harris.

and was the family's favorite child. Dane was very upset at the dog's death and blamed Nikola. A short time later, Dane suffered an accident (a fall, either from a horse or down cellar stairs) and died from his injuries. Nikola thought his parents blamed him for his brother's accident and worked hard to try to make amends. Nikola was a sickly child, and, destined for the ministry, he became seriously ill several times. As he recuperated from one such illness, he read Mark Twain's *Innocents Abroad*. In his autobiography, Tesla says it lifted his spirits enough to recover and started him thinking about America. Just after high school graduation, he contracted a serious case of cholera. As he was about to breathe what appeared to be his last breath, he revealed to his father that he hated the clergy and really wanted to be an electrical engineer. His father promised to send him to the best school if he would just recover. He did, and his father made good on his promise.

Tesla attended the Austrian Polytechnic in Graz, and spent many twenty-hour sessions studying electrical engineering. He irritated some of his professors by advancing beyond their knowledge. In an electric motor, a piece of equipment called a commutator was used to force the current to flow only one way and keep the motor turning in one direction. The commutator required contact between rotating parts, leading to frictional inefficiencies. Tesla asked why the commutator couldn't be eliminated. Despite ridicule from the instructor about the current flowing both ways—making it an oscillator rather than a motor—Tesla regarded the idea of designing a motor that had alternating currents a personal challenge. After more studies at the University of Prague (stopping short of achieving a degree), Tesla, in 1881, became the chief electrician for American Telephone and Telegraph Company in Budapest, Hungary. There, he suffered another major illness, which made his senses abnormally keen. Tesla said he could hear a watch ticking three rooms away.

During his recovery, the solution to the problem of the motor with current flowing both ways occurred to him. What he hadn't been able to figure out earlier was that the motor required a rotating magnetic field. The whole design burst upon him complete in all details. What he needed next was a working model, but that would take a while. The telephone station in Budapest was sold, and Tesla went to work for Continental Edison in Paris, where he became the company's troubleshooter. He thrived on long hours of work, rising at 5:00 am for a swim in the Seine, a stroll, and breakfast. He then arrived at the office by 8:30 and usually worked till late evening. After work, Tesla often ate at fancy restaurants, picking up the tab for whoever dined with him. He dressed in elegant clothes and cut a fine figure with his tall, slim frame and piercing blue eyes.

Continental Edison suffered an embarrassing problem (a short circuit blew out a wall) during the dedication of their electrical system at the railway station in Strasbourg. Tesla was sent to patch things up. Since fixing their public failure was important to the company, Tesla was promised a bonus if the repairs were completed quickly and well. While bureaucratic forms delayed the actual repair for days, Tesla found time to work on his AC motor project. He rented a machine shop near his hotel where the necessary parts could be made. There, Tesla built a working model of his new motor. Finally, both projects came together—the repair of the railway electrical system was complete, and his alternating current motor was finished. His Strasbourg friends, including the former mayor, were quite pleased with the repairs to the electrical system, but they were much less enthusiastic about his alternating current motor. When he returned to Paris after completing his assignment, his superiors were similarly uninterested in the motor project. They even failed to pay the bonus for the electrical system repair. They pointed out that very wealthy people had invested heavily in direct current systems and wouldn't be interested in generators that used different means. Tesla's boss suggested he go to America to present his idea to Edison himself.

Tesla quit his job, bought a steamship ticket, and gathered his meager possessions for the trip. As he boarded the train in Paris to begin the journey, his luggage was stolen. Thinking quickly, he boarded the train anyway, spending almost all his pocket money in the process. When he arrived at the embarkation point, he explained the situation to the boat officials. Tesla told them the time and location where he had bought the ticket and the number on it. He reasoned that either the thief would show up with the ticket, or no one would arrive, and he should be allowed to board. Although they were skeptical, the officials agreed to wait. When it was time to sail, and no one had appeared with the ticket, they allowed Tesla to be the last passenger aboard. After changing ships and enduring a rough crossing, Tesla arrived in America in 1884 with the equivalent of 4 cents, the least recorded amount for any immigrant at Ellis Island.

CONFLICT BEGINS

Tesla soon met Edison, who hired him on the spot. Tesla worked for Edison from the summer of 1884 through the spring of 1885, helping to trouble-shoot Edison's direct-current electrical systems. Edison was completely uninterested in Tesla's alternating current ideas, since he himself was so thoroughly committed to direct current. Besides, Edison told Tesla, alternating current is "a deadly current, whereas direct current is safe."[4]

In addition to the usual difficulties of selling and installing new technology, the price of copper became artificially inflated because speculators were attempting to corner the market. Edison's financial backers were displeased by the slow (in their view) progress and meager profits. Further, Edison had competition, as we'll see shortly. By 1887, Edison was feeling squeezed from several directions.

Edison put Tesla to work redesigning the company's direct-current generators to minimize losses. Tesla said that Edison offered him $50,000 to increase the generators' efficiency. Tesla worked with characteristic vigor—once for eighty-four hours straight—and completed the improvements Edison wanted. When Tesla tried to collect the bonus, Edison reportedly said, "Tesla, you just don't understand American humor." Tesla resigned in disgust.

Next, he formed his own company, Tesla Electric Light & Manufacturing. He satisfied his investors by designing an arc lamp for street lighting and industrial use, but the investors forced him out of his own company when he tried to develop the brushless alternating current motor he had designed in Strasbourg.

In 1887 and 1888, Tesla had to dig ditches to make a living, but a sympathetic foreman took him to see a Western Union official, who rounded up new financial backers. They organized the Tesla Electric Company, where Tesla was able to work out the mathematical details and build many different dynamos and motors, including a mechanical oscillator that shook neighboring buildings. In 1890, Tesla was invited to give a lecture to the American Institute of Electrical Engineers, titled "A New System of Alternating Current Motors and Transformers." He became an instant celebrity and accepted a one-million-dollar-plus-royalties offer for his patents. So, who could afford to offer a cool million? Read on.

PARTICIPANT 3: GEORGE WESTINGHOUSE

George Westinghouse (1846–1914) was born in the small village of Central Bridge, New York. His father manufactured farm implements, which exposed young Westinghouse to machinery at an early age. He

and his two brothers served in the Union military during the Civil War. After the war ended, Westinghouse spent a short time at Union College studying engineering. He cut his studies short to become an inventor. In 1865, he obtained his first patent, for a rotary steam engine. The railroad industry caught his attention. He designed a device for getting derailed cars back on the track, another that pre-

George Westinghouse (1846–1914). Used with permission from Sidney Harris.

vented derailments at switches, and a fail-safe braking system that used compressed air. These were followed by an automatic signaling system, in which electricity was used to indicate the passage of trains. Westinghouse patented his inventions after seeing his first few creations stolen by unscrupulous railroad managers. He started many different companies to produce his inventions and guarded his patent rights fiercely. An innovative employer, Westinghouse paid his workers well, cut the workweek from six days to five and a half, was among the first to institute paid vacations and pensions, and hired the first female electrical engineer. When an exploratory oil well on his property in Pittsburgh produced a gusher of oil and gas, Westinghouse designed and built a distribution system so the gas could be reduced in pressure and safely piped to many homes.

CLASH 1: BUSINESS

In 1884, Westinghouse hired William H. Stanley Jr., an inventor and patent holder in his own right. Not long after, Westinghouse read about a transformer (then called a secondary generator) invented independently by Lucien Gaulard (1850–1888) and John Dixon Gibbs (1834–1912) that was used to step down high-voltage AC to lower voltages suitable for lighting. It occurred to Westinghouse that if high-voltage electricity could be stepped down to lower voltages for home use, that would be similar to his pressure reducer invention. For the transformers to work, this innovation necessitated an alternating current (AC) system, as opposed to Edison's direct current (DC) system currently in vogue. However, AC's major advantage occurred at high voltage. The higher the voltage, the less the current that flows. The net result was that transmission losses could be minimized by stepping the voltage up to higher levels for transmission, then stepping it down for home or industrial use.

Westinghouse bought US rights to the transformer, and Stanley proceeded to improve the design to make it commercially practical. Westinghouse thus entered the electrical business with an AC system in 1886. Edison's DC system had about a four-year head start, but that didn't deter

Westinghouse. Westinghouse's company lit several commercial establish-
ments along the main street in Great Barrington, Massachusetts, where
Stanley lived and maintained his lab. The AC generator for this system
was initially a European import, but Stanley built an improved one. From
the electric lighting customers' point of view, there was little difference
between the Westinghouse system and the rival Edison system, also in use
in Great Barrington in 1886. Both offered lighting at substantially similar
rates (Westinghouse did undercut Edison prices occasionally), but the
major differences were Edison's inability to transmit long distances, and
Westinghouse's lack of an AC motor for industrial customers. By 1887,
after one year in business, Westinghouse had sixty-eight AC systems built
or under contract, while Edison had 121. If only Westinghouse had an AC
motor, he knew that he could compete with Edison across the entire range
of electrical business, besides just lighting.

COOPERATION 1: TESLA/WESTINGHOUSE

In 1888, Westinghouse paid the fees to license Tesla's AC motor so he
could compete with Edison for industrial customers. Once the motors were
built, Westinghouse was able to field a complete lineup of products comparable to Edison's. The major difference was that Westinghouse's AC
system didn't require the generator to be within a mile of the customer's
location.

CLASH 2: MORE BUSINESS

The AC/DC competition took a strange turn of events in 1888. On the good
side, Edison opened a state-of-the-art lab in West Orange, New Jersey. On
the bad side, several people were killed in electrical accidents, mostly electrical company employees who failed to observe safety precautions. These
accidents were followed by a scathing letter to the editor of the *New York
Evening Post* about the dangers of AC and how the public was in "constant danger from sudden death" because of AC. The letter writer, Harold

P. Brown, a seemingly obscure New York engineer, recommended that AC above three hundred volts be outlawed in the interest of public safety. Truly, there was some danger involved, but it was mostly due to the huge number of overhead wires already in place. They had been strung willy-nilly for arc-lighting systems, telegraphs, stock tickers (including the Gold Exchange), and other private electrical systems. In Brooklyn, the baseball team was nicknamed the Dodgers because Brooklynites had to dodge not only the streetcars but also dangling electric wires.

It seemed like an opportunity was being seized to discredit Westinghouse's AC system, even though that system wasn't entirely to blame. Westinghouse understood this and sent a warm, friendly letter to Edison, proposing peace between the companies. Edison's answer was: "My laboratory work consumes the whole of my time and precludes my participation in directing the business policy."[5] Westinghouse felt as if he had no choice but to fight. He appeared before the New York City Board of Electrical Control and quoted impressive safety statistics that favored his AC system over Edison's DC. By the end of July, Brown struck again. He held a demonstration at Columbia College in which he subjected animals to various electric shocks, trying to demonstrate the danger of AC. When the audience realized what he was going to do, many people left the room. Finally, an agent for the American Society for the Prevention of Cruelty to Animals stood up and forbade Brown to execute any more animals. The hostile audience filed out, with a diatribe against AC bouncing off their departing backs. After several other animal execution demonstrations, Brown began to work on a larger target: electrical execution (electrocution) of convicted murderers with alternating current as a "quick, humane" form of capital punishment. Letters, testimony, legal actions, legal fees, and many attorney-billable hours flew back and forth. The point was that the DC forces wanted to portray AC as a "killing current" so the public would fear it.

HERE'S WHERE IT GETS UGLY

In August 1889, the *New York Sun* published an exposé of Harold P. Brown. Someone broke into his Wall Street office and stole forty-five letters that showed he was paid by the Edison Company and the Thomson-Houston Company. (Edison and Thomson-Houston merged under the control of J. P. Morgan to form General Electric, GE.) Nevertheless, bureaucratic wheels continued to grind onward, proceeding toward the inevitable conclusion. In August 1890, convicted felon William Kemmler was executed in the electric chair at Auburn Prison in New York. Driven by a generator voltage of 1,000 to 1,400 volts, alternating electrical current surged through Kemmler's body for seventeen seconds, and the prison doctor pronounced him dead. As other attending physicians examined Kemmler, his chest suddenly heaved up and down. Quickly, they reattached the electrodes and ran the current for several minutes. Kemmler was finally dispatched, but it was hardly neat or even humane.

CLASHES PLUS FORCED COOPERATION

The electric current war (and a bulb war) continued in a less bizarre fashion, with the battlefield shifted to financial, governmental, and courtroom venues with small victories for each side. The next big skirmish was Chicago World's Fair of 1893. Both Edison, now backed by financier J. P. Morgan and called General Electric (GE), and Westinghouse/Tesla made bids to light the fairgrounds at night. General Electric's first bid was $1.8 million, later amended to $554,000. Westinghouse won the competition with a bid of $399,000. Edison was so displeased that he refused to sell them GE bulbs, so Westinghouse invented more efficient ones. As the fair opened, President Grover Cleveland pushed a button, and almost one hundred thousand incandescent bulbs lit the buildings and grounds spectacularly. Fairgoers dubbed it "The City of Light." Tesla's AC system worked perfectly and demonstrated its safety and efficiency to twenty-seven million fairgoers.

The next key battle was fought at Niagara, New York. The Cataract Commission was set up to decide on the best method for extracting energy from the raging Niagara River. After keen (and some not-so-keen) analysis, the commission awarded one contract to Westinghouse's AC system to generate power at the dam, and one contract to Edison's GE to license Westinghouse/Tesla patents for AC transformers and distributions systems and so they could run AC transmission lines to Buffalo, New York, twenty-two miles away. Within a few years, the power grid was 80 percent AC, so Tesla/Westinghouse were definite winners of the "current war."

COOPERATION DISSOLVES INVESTORS' COLD FEET

Although Westinghouse/Tesla won the Niagara battle, Tesla suffered a crushing setback. His Houston Street lab was completely destroyed in a building fire. All his plans, equipment, and partially completed projects were ruined, and were not covered by insurance. Not long after, Westinghouse's financial backers questioned his royalty agreement with Tesla. When Westinghouse passed along their concerns, Tesla responded by tearing up the agreement. Although it was eventually estimated to have cost him around $10 million, Tesla was optimistic that he would make much more. Tesla's research interests included wireless power transmission (radio), remote control of mechanical devices (robotics), and single-node vacuum tubes (X-rays).

Bonus Material: Tesla/Edison Internet interview. See To Dig Deeper for details.

AFTERMATH

The principal participants in the AC/DC current war all moved on after the conflict and accomplished plenty. Let's look at each of them.

Edison never rested long. He returned to an earlier interest and upgraded a large ore-crushing mill in Ogdensburg, New Jersey, that he had built earlier. His plan was to crush the ore so finely that electromagnets could separate the iron from the ore. He said, "I'm going to do something now so different and so much bigger than anything I've ever done before; people will forget that my name ever was connected with anything electrical."[6] Technical problems and a soft market made the business a failure, so Edison shut it down. However, one of the manufacturing by-products was sand that he sold to cement manufacturers. It turned out that the sand was of such a high quality that it made excellent cement. So, naturally, Edison next went into the cement business.

Edison's winter retreat in Ft. Myers, Florida, was quite near one of industrialist Henry Ford's homes. The two of them became personal friends and went on camping trips with businessman Harvey Firestone. All in all, Edison remained an inventor until the end. He held 1,093 US patents and 2,332 worldwide. Edison died in 1931 in West Orange, New Jersey.

George Westinghouse returned to an old interest of his: steam turbines. He acquired some patent rights and scaled up an earlier design to such a large size and high speed that it could power dynamos and propel ships. Westinghouse got his first patent at age nineteen and amassed 361 US patents. By 1900, Westinghouse's various companies employed fifty thousand people and were valued at $120 million. In the financial panic of 1907, Westinghouse resigned from all his companies. His health began to deteriorate in 1910, and he died in 1914.

Nikola Tesla had a new lab built in Colorado Springs in 1899, where he experimented with high-frequency, high-voltage systems, cosmic rays, atmospheric electricity, and electric oscillations of the entire earth/ionosphere system. Just after he started up his "magnifying transmitter," the power demands knocked out the Colorado Springs Electric Company's generator. Tesla's facility was soon closed and dismantled to pay debts. Tesla returned to New York and obtained financial backing from J. P. Morgan to build a laboratory on Long Island, called Wardenclyffe. The Wardenclyffe Tower was designed to be the hub of a world radio and wireless power distribution system. Tesla's plan was to shoot electrical currents (similar to lightning) into the upper atmosphere. A distant receiving station would conduct and distribute the current. The current would then flow into the ground, over a hundred meters deep, and complete the circuit by flowing back to the distribution station at Wardenclyffe. In Tesla's view, the earth and its ionosphere constituted a giant capacitor, with the rest of the atmosphere acting as a dielectric. Delays and cost overruns required more funds, and when Morgan found out that Tesla's ultimate aim was the free distribution of wireless power, he withdrew funding and the project collapsed. Tesla was severely discouraged, and his mental state may have deteriorated. He became more secretive than ever, lived in a succession of hotels (The Waldorf-Astoria, the St. Regis, the Governor Clinton, and the New Yorker), became morbidly afraid of germs (eighteen clean towels per day and dozens of napkins at each meal), and periodically issued grand statements about his current research, but without any evidence. In retrospect, some think he had obsessive-compulsive disorder (OCD). His reputation suffered, and eventually few took him seriously. He died in 1943 and was cremated shortly thereafter. He is still in the public eye, thanks to an opera about him called *Velvet Fire* that opened in 2004. Tesla is depicted as a modern Prometheus, who suffered mightily because he stole lightning from the gods. There is also a rock group named after him and a very successful electric automobile company.

Perhaps due to the catastrophic fire of 1893, or because of the many business failures, or even due to his secretive nature, Tesla wrote down precious little. He so captured the imagination of the world that many speculations and folk legends have sprung up about yet unknown Tesla ideas and inventions. The web is rife with speculations and rumors, but so far, there is no evidence to support them.

For chapter 8, we're going to return to earth, with a thud. An attractive theory gets no experimental support for the longest time, then . . . watch.

CHAPTER 8

ALFRED WEGENER STANDS HIS GROUND ABOUT CONTINENTAL DRIFT

My husband recently made me try on a bikini. A bikini is not so much as a garment as a cloth-based reminder that your parts have been migrating all these years. My waist, I realized that day in the dressing room, has completely disappeared beneath my rib cage, which now rests directly on my hips. I'm exhibiting continental drift in reverse.

—Mary Roach, American humor and popular science writer[1]

Maps are fascinating, little ones, big ones, ones that are a challenge to fold, whatever. Both of us (AW and CW) have been

World map. From Wikimedia Commons, user Saperaud~commonswiki.

map guys for a long time, and we've got plenty of company. The best maps are the ones that show the whole world.

The creator of the first modern atlas was the Flemish cartographer Abraham Ortelius (1527–1598), who started as a map engraver and illuminator, but with encouragement from the eminent cartographer Gerardus Mercator (1512–1594), Ortelius became a full-fledged geographer. After extensive travel and consultation with many other geographers, he issued the first collection of maps of the entire world in 1570, called an atlas, and published by Mercator. He continued to update this work, the last and most complete one being issued in 1597.

Even though this atlas was created prior to the Scientific Revolution of the later 1600s, Ortelius's world map was in itself a substantial observation. So, in the scientific method sense, Ortelius was a keen observer. And yet, mere observation wasn't enough for him. Look at the map again.

It's hard *not* to form a hypothesis similar to what Ortelius made in his book *Thesaurus Geographicus* (*The Geographic Thesaurus*): the Americas were "torn away from Europe and Africa . . . by earthquakes and floods."[2] Ortelius then continued: "The vestiges of the rupture reveal themselves, if someone brings forward a map of the world and considers carefully the coasts of the three [continents]."[3]

World map jigsaw puzzles intended for tots have occasioned many grown-ups (and kids, too) to arrive at a similar conjecture when they fit South America's east coast to Africa's west coast with such ease.

Over the next three hundred years or more, many others shared Ortelius's surmise to varying degrees, but let's skip ahead to 1912 to look at someone who stated this hypothesis much more strongly and gave it a name by doing so: continental drift. Meteorologist Alfred Wegener really stirred up a hornet's nest.

Alfred Wegener (1880–1930). From Wikimedia Commons, user Woudloper.

Alfred Wegener was born in Berlin in 1880, the youngest of five children. His father, Richard, was a theologian and taught classical languages at a local gymnasium. Alfred was first in his gymnasium class, then studied physics, meteorology, and astronomy in Berlin, Heidelberg, and Innsbruck. Although his 1905 PhD was in astronomy, he maintained a strong interest in meteorology and climatology. He worked with his older brother Kurt and carried out meteorological measurements, once setting a world's record in 1906 for continuous balloon flight of fifty-two and a half hours aloft. During the same year, Wegener participated in the first of his expeditions to Greenland, making more meteorological measurements. After his return, he became a lecturer in meteorology, applied astronomy, and cosmic physics at the University of Marburg. While there, he wrote a book titled *Thermodynamik der Atmosphäre* (*Thermodynamics of the Atmosphere*), that became a standard textbook.

In late 1911, Wegener happened across a scientific paper listing identical fossils of plants and animals on opposite sides of the Atlantic Ocean. After a bit of a search, he found many other similar cases of matching plants and animals. The standard explanation for these similarities was that in the past there were land bridges between continents, but that these bridges were now sunken below the ocean. It seemed far simpler to Wegener that the continents were once joined but had moved apart. Wegener found additional evidence for his hypothesis: some large-scale geological formations from separate continents matched closely, and fossils of tropical plants were found on Arctic islands.

Wegener first mentioned his "continental drift" hypothesis in lectures in 1912. He published a book titled *Die Entstehung der Kontinente und Ozeane* (*The Origin of Continents and Oceans*), which gave much more detail about his ideas in 1915. The hypothesis conjectured that about three hundred million years ago, all the lands were joined in a single supercontinent, called Pangea (Greek for "all lands"). The continents then moved across the face of the earth gradually, with the oceans filling in the spaces between them.

Wegener's goal in publicizing his theory was to begin a thorough, open discussion of the possibilities, rather than simple acceptance of a radical new theory. But that was not to be the case.

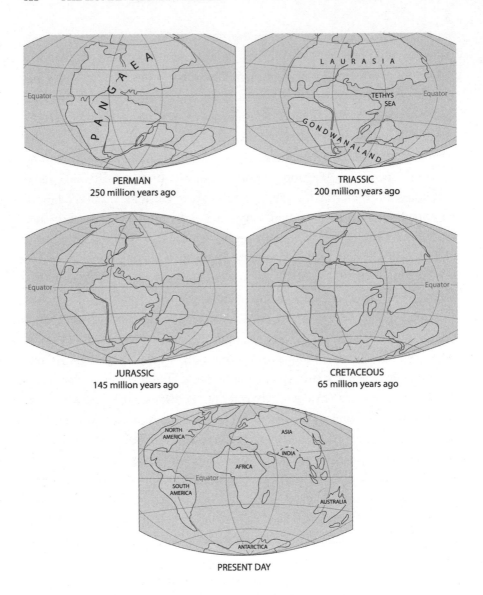

Continental drift. From the US Geological Survey.

BIG TROUBLE

Although there was some support, principally by South African geologist Alexander du Toit, British geologist Arthur Holmes, and Swiss geologist Émile Argand, the major reaction by geologists was swift and almost uniformly hostile. Rollin T. Chamberlin, geology professor at the University of Chicago, said, "Wegener's hypothesis in general is of the footloose type, in that it takes considerable liberty with our globe, and is less bound by restrictions or tied down by awkward, ugly facts than most of its rival theories." Chamberlain also suggested that "if we are to believe in Wegener's hypothesis we must forget everything which has been learned in the past 70 years and start all over again." In 1926, a major conference was held by the American Association of Petroleum Geologists to criticize the theory.

Why such outcry? First of all, Wegener wasn't trained as a geologist. He was considered by geologists to be an outsider or even an amateur. The fact that he was German and his writings didn't translate well into other languages didn't help in the time leading up to World War I. In addition, Wegener's idea was often misinterpreted as referring to the fit of the coastlines as opposed to the continental shelf, which is not subject to the same erosion and variation in the hardness or softness of rocks.

A bigger problem was the lack of a mechanism for the drift. What force could account for continents "plowing around in the mantle," as one critic put it? Wegener had several candidates for this force, but they were quickly demolished by opponents. As Wegener himself put it, "The Newton of drift theory has not yet appeared."[4]

In 1930, Wegener died while on his fourth expedition to Greenland. Continental drift theory was quietly swept under the rug, much to the relief of the geologists supporting the more popular notion of an unchanging Earth. Some years later, Arthur Holmes speculated that the driving force was supplied by currents in the mantle, the layer below the crust, but it took physical evidence from an unexpected source to help Wegener's ideas toward acceptance.

SUPPORT FROM AN UNLIKELY SOURCE

Victor Vacquier Sr. (1907–2009). American Geophysical Union (AGU), courtesy AIP Emilio Segre Visual Archives.

Victor Vacquier Sr. (1907–2009) and his family escaped the Russian Civil War by taking a one-horse sleigh across the ice-covered Gulf of Finland to Helsinki in 1920. Eventually immigrating to the United States, Vacquier earned a BS in electrical engineering and an MS in physics both from the University of Wisconsin. While working for Gulf Research in the 1930s, Vacquier invented an instrument for measuring the strength and direction of magnetic fields. This device was called a fluxgate magnetometer and was extremely accurate as well as being light and rugged.

Initially, the fluxgate magnetometer was used for petroleum exploration, but it also saw service in World War II as a submarine detector. After the war, in the International Geophysical Year (1957–1958), Vacquier directed a program from the Scripps Research Institute that used war surplus magnetometers to map the magnetic fields of the rocks on the ocean floor. The result was quite surprising: On either side of the deepest part of the ocean, the magnetic fields in the rocks had an alternating striped pattern. The simplest explanation for this observation was that the ocean floor was spreading, and when molten rock from the mantle below came to the surface, the magnetic ore (magnetite) in the molten rocks aligned itself with the earth's magnetic field at the time. Since the earth's magnetic field is known to have reversed direction at irregular intervals averaging two hundred thousand years, the pattern of reversals was frozen into the rocks on the ocean floor. Since continental drift could well have caused this pattern, Wegener's theory gained strong experimental support.

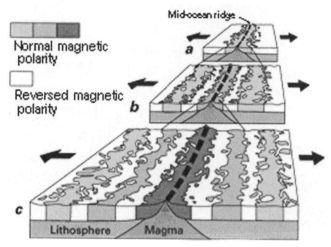

Magnetic stripes in ocean floor rocks. From the US Geological Survey.

These developments led to a new theory, championed by Canadian geologist J. Tuzo Wilson in 1965, called "plate tectonics" (from the Latin *tectonicus*, which translates to "pertaining to building," referring to the earth's crust).

The earth consists of a series of layers, which are, from the outermost to the innermost:

The Crust: a relatively thin, rigid layer made of fairly low-density rock and consisting of about a dozen major plates and many minor ones. The plates move slowly, dragged along by currents in the next layer down and cause volcanoes, earthquakes, and mountain formations as they collide and move apart.

The Mantle: the thickest layer made of higher-density rock that is very hot. At depths below the uppermost mantle, this layer flows quite slowly, like a glacier.

The Outer Core: a hot liquid layer of iron and nickel that is more dense than the mantle and sloshes around because of the earth's rotation.

The Inner Core: the earth's center, made of iron that is too hot and under too much pressure to be liquid.

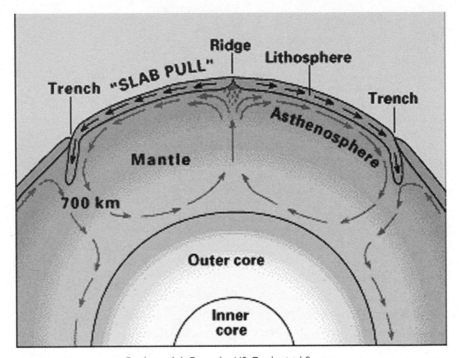

Earth model. From the US Geological Survey.

The plate tectonics model differs from Wegener's continental drift hypothesis in that the continents don't plow through the mantle; instead, the continents are part of the plates, which ride on the mantle and move because of underlying mantle currents.

So, when you look at that map we discussed earlier, realize that you are looking only at Earth's surface and think about all the activity in its mantle and cores below. And, if your next visit to Earth is a million years from now, your old map will be obsolete.

" I WOULDN'T WORRY. WITH CONTINENTAL DRIFT, AFRICA OR SOUTH AMERICA SHOULD COME BY EVENTUALLY. "

Used with permission from Sidney Harris.

Used with permission from Sidney Harris.

*"As far as continental drift is concerned,
it's a whole new ball game."*

Used with permission from Sidney Harris.

From the stable perspective of our lovely Planet Earth, let us head to the great beyond, starting with a seemingly well-grounded fellow: Albert Einstein.

CHAPTER 9

PART 1: ALBERT EINSTEIN, MARCEL GROSSMANN, MILEVA MARIĆ, AND MICHELE BESSO STRUGGLE WITH RELATIVITY

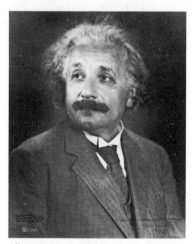

Albert Einstein (1879–1955). Image courtesy of the Observatories of the Carnegie Institution for Science Collection at the Huntington Library, San Marino, California.

Albert Einstein (1879–1955), *Time* magazine's Person of the Century, certainly set the tone for twentieth-century physics. This honor was bestowed on the grounds that he was preeminent among the many scientists in a century dominated by science. *Time*'s editors believed that the twentieth century "will be remembered foremost for its science and technology,"[1] and that their awardee, Albert Einstein, "serves as a symbol of all the scientists—such as Heisenberg, Bohr, Richard Feynman . . . who built upon his work."[2] Interestingly, the vast majority of Einstein's work was accomplished early in the century—from 1905 to 1918—so his scientific colleagues had plenty of time to surpass him—but none did.

Certainly his scientific prowess was amazing, but Einstein's influence goes way beyond science. He has been described as Charlie Chaplinesque, a cartoonist's dream with a recognizable face, mustache and hair, a quote machine, an internationalist, a pacifist, and a womanizer . . . in short, his humanity shines brightly.

In this chapter and the next, we will explore Albert Einstein's human activities and interactions, many of which informed and influenced his revolutionary scientific ideas, and others that are just, well, eccentric. He's definitely someone who would rank high on any "What historical person would you like to go to dinner with?" competition.

THE FAMILY

Hermann Einstein (1847–1902). From Wikimedia Commons, user Materialscientist.

Pauline Koch Einstein (1858–1920). From Wikimedia Commons, user Materialscientist.

Although Albert Einstein's father, Hermann (1847–1902), looks severe in his pictures, he was anything but. As a youth in Stuttgart, he enjoyed mathematics, but family finances were such that he needed a more practical education, and so he apprenticed as a merchant. He was a mild fellow who tended to consider all options very carefully before making any decisions. For a while, he sold feather beds in Ulm, making many friends but little money. Then, he married up.

Pauline Koch (1858–1920) came from a well-to-do family of corn merchants from Cannstatt, Württemberg. She was quiet, well educated, and had an inclination toward the arts, especially music. Pauline married Hermann Einstein when she was eighteen years old. The couple lived in Ulm, in a region called Swabia, where the Rhine and Danube Rivers are only a few miles apart and flow in opposite directions.

Before she turned twenty-one, Pauline

borc their first child, Albert, on March 14, 1879. Albert's head was overly large and misshapen at birth, causing concern for his parents, but it soon assumed a normal shape. Albert didn't please his Grandmother Koch, though. She proclaimed him "fat, much too fat."[3]

The earliest known picture of Albert Einstein shows a nonchalant fellow of age three. He had been very quiet, not even attempting to speak words until he was well past two. However, one of his early exclamations was quite memorable. When his sister, Maja, was born before Albert was three, he had evidently been promised a new toy. When his parents brought his sister home, Albert asked, "Yes, but where does it have its small wheels?"[4]

Maja Einstein (1881–1951) was born in Munich, where the family had moved when Hermann left the featherbed business for something more modern. Hermann and his brother Jakob had started an electrical business. The Einstein Brothers (their company name) manufactured direct-current dynamos and meters, and their major success was to supply lighting for the 1885 Munich Oktoberfest. Their chief competitor, Siemens's alternating-current system, soon created business problems for the Einsteins as Tesla/Westinghouse's efforts had done for Edison in the United States (recall chapter 7).

Albert Einstein at age three. From Wikimedia Commons, user Craigboy.

Maja and Albert Einstein, 1886. From Wikimedia Commons, user Materialscientist.

EARLY YEARS

Albert Einstein in Munich, 1893.
From Wikimedia Commons, user Quibik.

While they lived in Munich, the Einsteins shared a big house with Uncle Jakob and his family. They had a garden, flowers, trees and a lake nearby. Their house was an active place, with many relatives coming and going, one of whom was Elsa Einstein, three years Albert's senior. Elsa's mother and Albert's mother were sisters, and her father and Albert's father were first cousins, making her a cousin on both sides. Albert's first years of school were at a Catholic school. He called the teachers "sergeants" who rapped the knuckles of anyone who didn't answer imme-diately. Violin lessons started at age five, and young Albert showed a bit of temper by throwing a chair at one of his teachers. Maja Einstein said Albert's anger got so bad that his face went past red to yellow.[5] Also, he threw a bowling ball at her and hit her with a hoe. Nevertheless, Maja and Albert used to play duets together on piano and violin, with Mama joining in occasionally.

After elementary school, Einstein graduated to the gymnasium (public school), where he referred to the teachers one rank higher as lieutenants. Fortunately, something was always going on at home, including many cousin visits. And Albert enjoyed building houses of cards, some many stories tall.

As Einstein became a teenager, one of Germany's requirements needed to be fulfilled: every family was required to provide some form of religious instruction. Since the Einsteins were culturally Jewish but nonpracticing, they asked one of the more religious Koch cousins to help Albert learn enough about Judaism to fulfill the requirements. After studying the Bible and the Talmud (a central text of Rabbinic Judaism), Albert thought the ideas were beautiful and made up little songs of praise to God. He even stopped eating pork, although the family made fun of him. He began to prepare for his bar mitzvah.

And then an ironic thing happened. The Einsteins had been entertaining a struggling Polish medical student with Thursday night dinners. Although this fellow was ten years older than Albert, he brought interesting books for Albert to read. Albert devoured the books, and the two of them became friends. The books included works by Euclid, philosophers Immanuel Kant and Benedict Spinoza and even the book *Popular Books on Natural Science* by Aaron Bernstein, who was a kind of nineteenth-century Carl Sagan. The religion/science collision in Albert's mind was monumental. In his words, "Through the reading of popular scientific books I soon reached the conviction that much in the stories of the Bible could not be true. The consequence was a positively fanatic orgy of freethinking coupled with the impression that youth is intentionally being deceived by the state through

lies; it was a crushing impression."[6] The irony was that the medical student's name was Max Talmud. Later, after he completed his medical education, Talmud became a general practitioner in New York and changed his name to Max Talmey.

MOVING TO ITALY

In 1894, the Einstein brothers moved their electrical business to Pavia, Italy. Their families followed shortly—all except Albert. He was left in Munich to finish his

Max Talmud (1869–1941). Photographed by E. Mulhern, c.1890.

schooling at the gymnasium with the "lieutenants." It must have pained his mother sorely, as it must have disappointed Albert also. But he put up a good front in his letters. Albert stuck it out for six months, then got a family doctor friend to diagnose him with "neurasthenic exhaustion,"[7] a condition that might be provoked by a few sleepless nights.[8] Albert hopped a third-class train for Italy and surprised his family. He arrived with a two-pronged plan: He would study and take the entrance exam for the Federal Swiss Polytechnic School (ETH), which prepares its graduates for a career in teaching, and he would renounce his German citizenship, which would free him from the compulsory military duty at age seventeen.

The family bought it. What else could they do? Sure enough Albert studied, took the entrance exam for ETH, sure enough Albert passed the physics and mathematics parts of the exam with flying colors, and sure enough he failed the history and language parts of the exam. Although the physics professor, Heinrich Weber, graciously told Albert he was welcome to sit in any of his classes whenever he was in Zurich, the Polytechnic director, Albin Herzog, suggested that Albert complete secondary school before entering the ETH.

The place Herzog sent him was only thirty minutes away from Zurich, in the pleasant little town of Aarau on the Aare River. Albert was to board with the school's director, Jost Winteler. This whole experience turned out to be a boon for Albert on several fronts. The Wintelers had seven children, and they were a boisterous lot. The situation was similar to Albert's home in Munich but better. There were parties, kite-flying sessions, hikes in the hills, and endless discussions about everything. Albert, the foreigner, had slightly different ideas and loved to pontificate about big issues, often dressed only in his bathrobe.

He became a gregarious teenager. Classes at the school were far more free-wheeling than the rigid Germanic ones Albert was used to, and he became a freethinker. His thoughts were often about deep concepts in physics. Conducting thought experiments was well known from ancient Greek times. The German term for is *Gedankenexperiments*. For example, Albert tried to visualize what it would be like if someone moved at the speed of light and observed another light wave traveling alongside. Would it be a frozen light beam? That didn't seem right to him. In later life, he considered such thoughts as the beginnings of his ideas about relativity.

FIRST ROMANCE

The youngest Winteler child, Marie, was two years older than Albert, and very attractive. The two of them became quite an item, with Marie even exchanging letters with Albert's mother. Classes went well, and Albert completed high school in 1896 at the top of his class.

The idyllic time in Aarau ended with Marie taking a teaching job at Olsberg and Albert going off to Zurich to attend (finally) the ETH, or Poly, as it was called. Marie offered to still do his laundry. Albert accepted.

In Zurich, along the banks of the Limmat River, the ETH was colocated with the prestigious University of Zurich. Six new students were added to the ETH's Section VI A, physics and mathematics, for the class of

Einstein's high school graduating class in Aarau, 1896.
Used with permission from HIP/Art Resource, NY.

1900. One was Albert Einstein. Two of the others exerted a profound influence on Albert's life. Besides Albert's classmates, two others with ETH connections also impacted his life enormously. One was a helper all the way, and the other became disillusioned and set out to torpedo Albert's career but instead accomplished the exact opposite. Let's start with the positive one, classmate Marcel Grossmann.

GOOD FRIEND

Marcel Grossmann (1878–1936).
From Wikimedia Commons, user Melirius.

Marcel Grossmann's father owned an agricultural machinery factory in Budapest. After Marcel completed gymnasium there, the family moved to Basel, where he finished high school, then prepared to be a mathematics teacher by enrolling at the ETH. Marcel was a keen judge of character and quickly made a friend of Albert Einstein. Taking him home for dinner, Marcel announced to his parents, "This Einstein will one day be a very great man."[9] At the ETH, Marcel and Albert took many of the same classes. Marcel always attended and took careful, detailed notes. Albert, on the other hand, skipped many classes, especially mathematics. He was busy exploring physics topics that interested him. Although these extracurricular topics were most interesting to Albert, many discussions with Marcel over iced coffees in the cafes of Bahnhofstrasse served to keep these ideas in perspective. Marcel turned out to be far more helpful than just a coffee-drinking buddy.

Albert Einstein made another lifelong friend in 1896: Michelangelo Besso.

Michelangelo, or Michele, as Einstein called him, was six years older and an engineer who had gotten his degree from the ETH. Both being violin players, they met at one of the evening musical gatherings that Albert frequented. Besso was short, with dark curly hair. He was gregarious and full of curiosity, especially about the philosophy of physics. Albert and Michele became good friends almost immediately.

Anna Winteler Besso (1872–1944) and Michelangelo Besso (1873–1955). Besso Family, courtesy AIP Emilio Segre Visual Archives.

Because of Besso's short stature, he had trouble meeting eligible women. Albert provided a good solution. He introduced Michele to Anna Winteler, Marie's oldest sister, who was even shorter. They fell in love and married within two years. In a letter, Albert described Besso fondly: "He has an extraordinarily keen mind, the disorderly workings of which I observe with great enjoyment."[10]

SERIOUS ROMANCE

Mileva Marić (1875–1948). From Wikimedia Commons, user Kelson.

The other addition to the class of 1900 was an unusual one: a woman named Mileva Marić. Mileva was Serbian, the daughter of a wealthy family from Titel in the Austro-Hungarian monarchy. She was born with a dislocated left hip and walked with a limp. Her mind was quick, and she excelled at mathematics and physics to the point that her father had her admitted to all-male schools so she could progress. Mileva and Albert circled each other warily and connected only superficially during the 1896–97 school year, except for a pleasant hike in July 1897. In the fall of 1897, instead of returning to the ETH, Mileva went to study at the University of Heidelberg. Albert and Mileva did not correspond often, but when they did, it was usually about physics or gossip. His letters were addressed to "Fraulein Marić," and hers to "Herr Einstein." While at Heidelberg, Mileva studied under Professor Phillip Lenard, a noted experimentalist. She dutifully reported to "Herr Einstein" about Lenard's recent investigations with cathode rays, which were later realized to be electrons. Shining light on metals causes cathode rays to be emitted, with a strange connection to the wavelength of the light. Shorter wavelengths of light caused the electrons to have more energy, and above a particular wavelength, no electrons were given off at all. This was called the photoelectric effect. Lenard also lectured about another topic of great interest to Einstein: Brownian motion—the little dance executed by tiny particles viewed through a microscope. Lenard's results, and Lenard himself, eventually played a larger role in Einstein's life—but we'll get to that later.

Albert told Mileva about enjoying Professor Weber's physics lectures, which were delivered "with great mastery."[11] He said he had taken copious notes that he was willing to share with Mileva when—and if—she returned. Return she did, in April 1898. Yes, Albert and Mileva did share a lot of notes, including some of Marcel Grossmann's, but a watershed had been reached. At the midway point of Professor Weber's year-long extensive course titled "Physics," Albert realized that James Clerk Maxwell's electromagnetism and Ludwig Boltzmann's statistical mechanics were not going to be discussed. This omission of recent works angered Albert greatly. Weber seemed to be rooted in the past, so Albert and his study buddy Mileva began reading the modern works of Boltzmann, Paul Drude, Ernst Mach, Hermann Helmholtz, Heinrich Hertz, and Henri Poincaré. Albert began skipping Weber's lecture classes, and even when he did attend, he caused great annoyance by disrespectfully addressing Herr Professor Weber as Herr Weber.

In the fall of 1898, intermediate exams were given to ensure students were making satisfactory progress toward the second half of the program, when the final exams would determine a student's eligibility for a diploma. Thanks to all his class skipping, Albert had few notes to study to prepare for the exam. Marcel Grossmann rescued him by offering his notes for Albert to study. As distasteful as Albert found cramming, he had little choice, so he buckled down and studied hard. The exams were oral, and the results quite remarkable—Albert scored 5.7 out of 6, higher than Marcel, who scored 5.6. This was save number one for Grossmann. Mileva postponed the exam for a year.

ACADEMIC STRUGGLES

Following the summer break, the school year progressed, and the friendship between Albert and Mileva progressed even faster. She began to call him Johnnie, he called her Dollie, and they studied Hertz and Helmholtz, feeding Albert's obsession about the æther—that medium postulated for carrying electromagnetic waves. Although Albert dutifully attended Herr Professor Weber's lab classes and enjoyed them, he skipped most lectures and the once-hopeful relationship between Albert and Weber continued to deteriorate. At one point, Weber told him: "You're a clever boy, Einstein, extremely clever, but you have one great fault. You'll never let yourself be told anything."[12]

Einstein also had a lab class with the other physics professor at Poly, Jean Pernet. Einstein disliked Pernet and the class. He would often toss the experiment instructions into the wastebasket and perform the lab according to the way he thought it should be done. Once, Einstein had an accident in lab that required stitches in his right hand, preventing him from playing his violin for several weeks. Pernet failed him and reported him to the dean, who reprimanded Einstein for "nondiligence."

Einstein spent most of the summer vacation with his mother and sister at the Hotel Pension Paradise in Mettmenstetten, only a short train ride from Zurich. His ever-present physics books took up much of the mornings, and in the afternoons he hiked and played the violin. The hotel owner's sister-in-law was a very attractive seventeen-year-old, Anna Schmid. Einstein enjoyed flirting and making music with her. His mother, who disapproved of his growing attachment to Mileva, probably welcomed his attentions to another woman. Einstein wrote a pleasant verse in Anna's autograph book at summer's end:

> You girl small and fine
> What should I inscribe for you here?
> I could think of many a thing
> Including also a kiss
> On the tiny little mouth.
> If you're angry about it
> Do not start to cry
> The best punishment is—
> To give me one too.[13]

FINAL YEAR AT THE ETH

In the fall of 1900, Mileva finally sat for her postponed intermediate exam and passed. During the remainder of this last year of their program, the established pattern continued: Einstein studied independently and skipped classes (except for Weber's lab, which he enjoyed). Mileva tried to catch up but was distracted by her Johnnie; and Marcel Grossmann attended all the classes, took great notes, and drank coffee with Einstein. This was their comfortable routine—with one exception—Dollie and Johnnie began to talk about marriage. Marie soon faded from Albert's life, replaced by his studies and a stronger female interest. At this point, it was just talk, because parental wishes were not likely to be favorable. But as spring drew to a close, the July final exam time loomed large. The established pattern from the intermediate exam was repeated: Einstein borrowed Grossmann's notes, crammed, then sat for the exam, which included both written and oral parts. This time, the results were a bit different. Einstein was fourth out of five and just barely passed. This got him his diploma, but just by the skin of his teeth.

Mileva wasn't so fortunate. She failed the exam, mostly because of her difficulties with math. After listening to Einstein's encouragement to study hard and take the exam again next year, she beat a hasty retreat to Novi Sad and her family. Einstein met his mother and sister for a small holiday in Melchtal, a small resort valley in Switzerland. Although he was warned by his sister that their mother would question him about what would become of Mileva, Einstein blurted out the exact answer she didn't want to hear: "My wife." His mother hurled herself on the bed and sobbed like a baby. Then she told Einstein what she thought of Mileva, omitting nothing. Einstein's temper flared, and he denied they had "been living in sin," but made no promises. Both sides retreated, erroneously believing they had made progress in changing the other's mind. After the holiday ended, Einstein returned home, helped his father in the electrical business for a while, and then returned to Zurich.

POSTGRADUATE DIFFICULTIES

Much to his astonishment, Einstein found that the other three class of 1900 graduates had all been tendered assistantships at the ETH, but he had no such offers. In fact, Professor Weber took on two electrical engineering graduates as assistants. Einstein applied to several other professors and nearby schools to no avail. He had no job and was running low on funds. Then, it dawned on him. Perhaps his relationship with Professor Weber had sunk to the point that Weber was trying to push him out of physics. Was Einstein's criticism of Weber's "old-school" attitude about theoretical physics so disturbing to Weber that he would give negative recommendations for Einstein's job applications? Did Weber agree with Professor Jean Pernet, who once asked Einstein, "Why don't you study medicine, law, or philology instead?"[14]

Heinrich Friedrich Weber (1843–1912) earned his doctorate in physics in 1865 and worked in several laboratories, with an emphasis on precise instrumentation as opposed to more theoretical pursuits. At the ETH, Weber focused primarily on building the physics laboratory program. He had few publications, and these were experimental determinations. Jean Pernet had been hired as Weber's assistant, but he became so dissatisfied with the position that the two barely spoke. Physics was not one of the top subjects at the ETH, but Weber was nonetheless determined to build a first-class lab. Perhaps a too-independent, disrespectful, even arrogant student such as Einstein annoyed him beyond his breaking point. In any event, he and Einstein disagreed about Einstein's proposed PhD thesis: Einstein wanted a theoretical topic, while Weber wanted him to pursue an experimental investigation.

Even though Einstein's lack of a job and disagreements with Weber made him wonder if Weber wasn't trying to dissuade him from physics, he was in fact driven to explore the subject more deeply. Einstein persuaded Professor Alfred Kleiner at the University of Zurich to accept him as a doctoral student. He did his best to minimize depressive thoughts about missing Mileva, his parents' negative attitude about her, his father's failing business, and his own job troubles. Like the proverbial ostrich with his head in the sand, Einstein buried himself in physics (and music). His mind was abuzz with ideas about unsettled areas of physics: æther, atoms, the quantum, thermodynamics, and even capillarity, the rising of liquids in thin tubes due to attractive forces between molecules.

HOPE

Finally, in 1901, the logjam broke. After a rare complaining letter to Marcel Grossmann, the return letter contained a ray of hope. Marcel said that his father was good friends with Friedrich Haller, the head of the Swiss Patent Office in Bern and said there might be a job there for Einstein as a patent examiner. It was not the academic job Einstein desired, but this was no time to be fussy. With the slow pace of grinding bureaucratic wheels, his hiring might take a while, but it seemed like a good possibility.

This job is save number two for Grossmann, considering the loan of his college notes save number one, and there is still more to come. The following day, another piece of good news arrived: A high school geometry teacher in neighboring Winterthur needed someone to teach for him while he fulfilled a military obligation, and he asked Einstein to help out.

Convinced that a corner had been turned, Einstein wrote to Mileva, inviting her to meet him at Italy's Lake Como. Einstein was confident that Mileva would pass her exams, he would have a job, and things would be much better. After some vacillation, she agreed, and they met on May 1, 1901. They had a lovely time. Afterward, she went to Zurich to complete her exam preparations, and he went to his temporary job in Winterthur. Einstein visited her in Zurich on Sundays, and indeed they were both in much better moods than before. On one of those Sunday visits, Mileva had some news to share: "I'm pregnant."

TROUBLES MULTIPLY

As if one disaster wasn't enough, Mileva failed her exams (again) and had to return to face her parents. Einstein wrote a letter to Mileva's father promising marriage, but that hardly mitigated her parents' reaction. Her second exam failure added to the problem, and a letter from Einstein's mother served as the icing on this inedible cake. Einstein's mother's exact words were not preserved, but her tone was clear, even though she was unaware of the impending baby. Both sets of parents were unwilling to see the lovebirds united, but the pregnancy had its own logic. Mileva stayed home and suffered through her parents' wrath while Einstein returned to his tutoring job near Winterthur to try to earn some money.

Later in the fall, Mileva stayed at Stein am Rhein so that Einstein could visit her on his days off. Einstein tutored and worked on his dissertation, which related to the kinetic theory of gases and intermolecular forces. In November 1901, a seven-month pregnant Mileva returned home to Novi Sad, and Einstein submitted his dissertation to Kleiner. Soon, an ad appeared in the *Federal Gazette* for a patent examiner at Bern. The qualifications appeared tailored to Einstein. Late in December, Einstein visited Kleiner at Zurich and was dismayed to find that he hadn't yet read Einstein's dissertation. Kleiner promised to do so over the Christmas break. They had a very productive discussion about the electrodynamics of moving bodies and Einstein's idea for an experiment to check the æther. Kleiner was very encouraging and offered to write letters of recommendation for Einstein. Early 1902 saw Einstein about to move to Bern in anticipation of the Patent Office job. On the way, he visited Professor Kleiner again, and Kleiner told him he needed to rewrite the thesis, so Einstein withdrew it.

The next big happening was a birth in Novi Sad. Mileva's father wrote a letter to Einstein. At first, Einstein feared the worst when he saw the letter from Mileva's father, but when he summoned the courage to open it, he was relieved to hear Mileva was frail but alive. Mileva had an extremely hard labor, which left her weak and exhausted. The child was a girl, named Lieserl, a diminutive for Elizabeth. Einstein's return letters were full of support and love, but his impoverished state made any offer of help or even a visit impossible. He needed money.

OLYMPIA ACADEMY

Upon arrival in Bern, Einstein advertised tutorial services in the newspaper but got only one taker, Maurice Solovine, a philosophy and physics major at the University of Bern who felt his scientific education was lacking. After a long conversation, Einstein abandoned his tutorial ideas and suggested they continue to meet and talk. Soon, he invited Conrad Habicht, a friend from high school days in Aarau to join their discussions. Although they gave themselves the name the "Olympia Academy," in a playful sense, they did read and discuss a wide variety of books, including ones by David Hume, Ernst Mach, Karl Pearson, John Stuart Mill, and Henri Poincaré. One of the hallmarks of this bohemian group was that its members sought wide-ranging general principles. They were after "the big picture," and as anyone who has taught knows well, the act of explanation clarifies the idea not only to the listener but also to the speaker.

Einstein's primary fascination was the statistical mechanics theories of Ludwig Boltzmann, which attempted to bridge the gap between the realm of the very small—atoms—and the large world of classical physics. He carricd on a correspondence with Paul Drude, the editor of *Annalen der Physik* (*Annals of Physics*) about what he saw as flaws in Drude's and Boltzmann's theories. He was also extremely interested in recent papers by Max Planck, who was the associate editor of the journal. Planck had proposed a mathematical solution to the problem of blackbody (ideal bodies that reflect nothing, only radiate) radiation that involved energy being arranged in packets or quanta (singular quantum). Einstein suspected a connection to the æther problem he had wrestled with for many years, but he couldn't quite put his finger on it.

A JOB

Finally, in late May 1902, he had his interview at the Patent Office. He got the job and started in June at the lowest rung on the ladder, that of technical expert third class. The pay, however, was twice what he would have made as an assistant at Poly. Patent Office director Friedrich Haller recognized Einstein's lack of education in blueprint reading and taught him that skill. Haller was gruff but straightforward, and Einstein appreciated his help.

Einstein was a fast learner, so he settled into the Patent Office routine in a matter of weeks. While the job required more thought than he anticipated, it appealed to his critical nature, and he considered it an interesting challenge. Soon, he was able to spend an occasional moment on the small sheets of material in his center desk drawer—his Department of Theoretical Physics, he called it. Foot-

Albert Einstein in the Patent Office, about 1904. © Underwood & Underwood/Corbis.

steps in the hallway outside his office caused the small sheets to disappear into the drawer in short order. Einstein's Patent Office job only reinforced his "outsider" status in the academic world. Accordingly, in his spare time, Einstein continued to search for a suitable thesis topic and wrote papers for publication in the *Annalen der Physik*. Papers were not peer reviewed but were published if accepted by the editor, Drude, or assistant editor, Planck. Einstein's first papers, published in 1902, 1903, and 1904, all dealt with generalizing the foundations of statistical mechanics, as originally set forth by Boltzmann. They were quite insightful but were substantially the same ideas as J. Willard Gibbs had published in America, prior to Einstein. Einstein didn't read English, so he had no knowledge of Gibbs's work.

DEATH AND MARRIAGE

Einstein's routine of patent examinations, theoretical physics work, Olympia Academy meetings, and sleep suffered a substantial interruption in early October in the form of a letter from Milan. His father was seriously ill. Einstein rushed home only to find his father on his deathbed. Just before he passed away, he gave Einstein permission to marry Mileva. Whatever second thoughts Einstein may have had about marriage dissipated, and they were married in a civil ceremony at Bern City Hall on January 6, 1903. The Olympia Academy was represented by Solovine and Habicht, who were the only witnesses. After the wedding and celebration at a local restaurant, the newlyweds returned home to a familiar scene. Einstein had forgotten his keys and had to wake the landlady to let them in—again, as it had happened many times before.

In the summer of 1903, Mileva returned to Novi Sad, probably to arrange for Lieserl's adoption. Whatever happened to Lieserl hasn't been discovered, but speculation is rampant. The fact is that all traces of her vanished, and Einstein never saw her. While there, Mileva learned that she was pregnant again and wrote of this development to Einstein, possibly fearing a negative reaction. "I'm not the least bit angry that Dollie is hatching a new chick. In fact, I'm happy about it" was his response.[15]

The Olympia Academy now had a new member, but Mileva added little to the discussions and didn't join the frequent hikes. She got along well with Solovine and Habicht, but perhaps she preferred quieter discussions with her husband. Additionally, Einstein submitted papers and book reviews to *Annalen der Physik*, which helped keep his physics up to date. In the fall of 1903, Habicht was offered a job, so it appeared the Olympia Academy would no longer be at full strength. Before long, there was an opening at the Patent Office, and Einstein encouraged his old friend Michele Besso to apply. He did, and he got the job. Another deep thinker was thus added, along with Besso's wife Anna (formerly Winteler), who became good friends with the quite-pregnant Mileva.

FAMILY RESTART

In May 1904, the newest Einstein was added: Hans Albert. He was a lovely baby with a marvelous disposition, and he brought great joy to his parents. Einstein was an attentive father and often sat with Hans Albert on one knee and a pad of physics equations on the other. Imagine the small Einstein apartment, filled with drying diapers, cooking smells, puffs of smoke, and happy baby sounds. Then think of Einstein himself, walking (with Besso) to his Patent Office job six days a week, examining patent applications (plus some work on his center drawer material). The busyness seems hardly conducive to serious scientific work, yet this was Einstein's peak period of productivity. Here's a view of Einstein's youthful routine from his sister: "His work habits were rather odd: even in a large, quite noisy group, he could withdraw to the sofa, take pen and paper in hand, set the inkstand precariously on the armrest, and lose himself so completely in a problem that the conversation of many voices stimulated rather than disturbed him; an indication of remarkable power of concentration."[16] So, the giant leap Einstein made in 1905 may have its roots in his ability to concentrate in the midst of chaos.

ANNUS MIRABILIS—THE MIRACLE YEAR

Several fundamental topics had been on Einstein's agenda for a while: atoms and molecules, electrodynamics and relative motion, and radiation and the quantum. Professor Kleiner had found the work on electrodynamics interesting, so Einstein wrote up what he had (although he considered it incomplete) and submitted it to Kleiner at the University of Zurich. Within weeks, Kleiner's rejection came in the mail. He had found the mathematics incomprehensible.

Although annoyed, Einstein pressed on with his other research. He took Planck's idea that blackbody radiation came in bundles called quanta and extended this notion to light in general. To support the idea of light having both particle and wave nature, Einstein explained the curious experimental result of Philip Lenard (who will show up later) from several years earlier. Called the photoelectric effect, this experiment shows that light incident on a metal causes the ejection of electrons, but the electrons' energy depends not on the light intensity but on its frequency. Einstein's explanation was that the light acted like a particle and gave its energy to the electron in a collision on an all-or-nothing basis. Since the light's energy is proportional to its frequency, the energy of the ejected electron also increases with the light's frequency. Einstein wrote these thoughts in the form of a paper and submitted it to *Annalen der Physik* on March 17. The paper was accepted without comment and published in June, even though the editor, Max Planck, didn't really like Einstein's extension of his original idea, or even take it seriously.

Soon after the photoelectric effect paper was in the mail, Einstein had a conversation with Besso about his thesis troubles. As Einstein poured sugar into his tea, he thought about how the tea's viscosity (think gumminess) changed when sugar molecules were introduced. If he could just relate the viscosity and the diffusion coefficient (think how fast the sweetness spreads) to the size of the sugar molecule, he could make an estimate of the molecule's size. Besso thought this sounded like a good idea, so Einstein wrote it up and submitted it to Kleiner. At first, Kleiner rejected it as being too short (seventeen pages), so Einstein added one sentence and resubmitted it. Kleiner looked at it more carefully and then had the mathematics department head check it. When it passed inspection, Kleiner informed Einstein that his thesis was accepted. Finally, he became Doctor Einstein. He submitted this work for publication in *Annalen der Physik*; wherein it was published about six months later.

Expanding his idea of molecules in constant random motion, Einstein reasoned that even large particles would undergo collisions with molecules, and the jittery motion of large particles would be detectable if the size of these particles was sufficient to appear in a microscope. Besso assured him that this phenomenon does indeed happen, and is called Brownian motion, after Scottish biologist Robert Brown, who had observed it some seventy-five years earlier. Einstein's analysis showed that a tiny sphere would move a distance of about one diameter in one second. Thus, although one couldn't see molecules directly, the observable motion of visible spheres reinforced the idea of molecules in constant random motion causing observable phenomena. He wrote this up and sent it to *Annalen der Physik* for publication.

Three papers down and one to go. Einstein's long wrestling match with the æther, that medium in which electromagnetic waves were thought to move, was next on his agenda. After a thorough day-long discussion with Besso about every aspect of the inconsistency of Maxwell's electrodynamics and Newton's mechanics, the two of them boiled this inconsistency down to a single question: How could there be an absolute reference frame for space and time and yet a constant speed for electromagnetic waves? Something had to give, and eventually they became tired. Einstein announced he was giving up on the quest, but, somehow, that evening it all came together in his mind. The following day, he announced to Besso, "Thank you. I've completely solved the problem."[17]

The solution was nothing short of revolutionary and took him six weeks to get it written up for publication. "On the Electrodynamics of Moving Bodies" is possibly the most famous paper in physics. It's usually called by its short name: special relativity. The æther was notoriously absent, as it was totally unnecessary. The only acknowledgment in the paper was: "In conclusion, let me note that my friend and colleague M. Besso steadfastly stood by me in my work on the problem here discussed, and that I am indebted to him for many a valuable suggestion."[18]

Some giant principle had to give, either absolute space and time or the constancy of the speed of light. Einstein chose to overturn Newton's absolute space and time, making them relative, depending on the motion of the observer, while keeping the speed of light constant. The paper was submitted in June and published in September. An exhausted Einstein took some time to rest and then produced another short paper, which led to the famous $E = mc^2$ result. Finally, he went on a short vacation to Mileva's home in Novi Sad so she could show off her handsome husband and new son, Hans Albert. Back home in Bern, Einstein waited for reaction to his monumental efforts. His 1905 papers established atoms, took light quanta seriously as a feature of nature, overturned absolute time and space, and demolished the concept of an æther. Who could ask for anything more? Still, as we know, theories are interesting, but experimental evidence was needed. There were a few requests for reprints of the papers, but these were dribbles rather than a torrent. A request for a reprint of the Brownian motion paper came from Heinrich Zangger, a professor at the University of Zurich's veterinary school. Einstein must have wondered about that one, but Zangger would eventually become a lifelong friend. Still no substantial recognition from the physics community. Einstein continued his day job at the Patent Office.

According to Einstein's theory of special relativity, not only do space and time measurements depend on the relative motion of the observer, so, too, does the mass of an object. Specifically, the apparent mass of a body increases as it moves faster, with a more noticeable effect as a body nears the speed of light. Several measurements of the apparent mass of electrons traveling at speeds near the speed of light yielded increased mass values, with the most accurate experiments conducted by Alfred Bucherer in 1907. Bucherer's results matched Einstein's prediction closely. Einstein's theory of special relativity was thus supported by experimental evidence, but he still toiled away at the Patent Office, far from his academic dreams. That was all about to change, and fairly soon.

CHAPTER 10

PART 2: ALBERT EINSTEIN'S STRUGGLES CONTINUE

Einstein's torrent of papers and book reviews continued past his monumental year of 1905. Fortunately, they generated attention in just the right place. His thesis adviser, Alfred Kleiner, had been the lone physicist on the staff of the University of Zurich for some time. He was finally coming into a spot in which he could get some help. He was about to become rector of the university, and he might be able to create another physics position. Although he had earlier encouraged Friedrich Adler to prepare for the position he was going to set up, Kleiner saw that Einstein's enthusiasm and theoretical interests would make him a better fit. Kleiner wrote Einstein and suggested he become a privatdozent (unsalaried lecturer) at the University of Bern as a first, preparatory step toward the eventual professorship at the University of Zurich. In short order, Einstein completed the necessary materials and was approved.

Since he still worked at the Patent Office, he could offer classes only at odd times. This cut into his valuable research time as he worked to extend his special relativity to cover accelerated motion. Only the tantalizing possibility of a full-time job at the University of Zurich kept him going. Meanwhile, home life with Mileva suffered some serious neglect.

Finally, after Einstein had delivered a pretty good lecture to the Physical Society of Zurich in

Alfred Kleiner (1849–1916).
From Wikimedia Commons, user
Tianxiaozhang~commonswiki.

February 1909, Kleiner was able to give Einstein a strong recommendation, and the committee voted to make him an offer. After an initial salary offer that Einstein thought was too low, the committee agreed to match Einstein's current remuneration. He accepted. Friedrich Haller was amazed when Einstein resigned from the Patent Office effective October 1909 to become a professor at the University of Zurich. Academe had beckoned, and Einstein heeded its call.

PERSONAL DISASTER

News of Einstein's acceptance of a position at the University of Zurich showed up in several newspapers and attracted the attention of Anna Schmid. Einstein had met Anna many years earlier when he was on holiday with his mother and sister at the Pension Paradise Hotel (see chapter 9 for his poem to her). Anna congratulated Einstein, told him of her marriage to George Meyer, and said they were living in Basel. Einstein wrote back and invited her to come see him in Zurich at the university. Anna responded positively, and her letter was somehow intercepted by Mileva. Immediately, Mileva became suspicious that her husband was having an affair and dashed off an angry note to Anna's husband, telling him how outraged she was that his wife and her husband were carrying on so. When Einstein found out about Mileva's letter, he was furious. He wrote a letter of apology to Meyer immediately, telling him he hadn't seen Anna in ten years, and nothing had happened between them even then. Somehow, the similar letter his mother had written to Mileva's parents must have come to his mind. Einstein wondered if either of those two primary women in his life trusted him. Lost trust is difficult to regain.

ACADEMIC STARTUP

Einstein began his academic career lecturing at the University of Zurich in October 1909. He and Mileva thought that perhaps another child might get their relationship back on track. Mileva became pregnant almost immediately. Although teaching made him a bit nervous at first, Einstein soon relaxed and became a much less formal professor than his colleagues, which pleased the students enormously. They surrounded him, even after class, following him to the cafés in Zurich for discussions of science. He became more confident but remained humble and friendly. Students and colleagues loved him. Mileva, however, was still getting the short end of the stick.

ACADEMIC STEPPING-STONE

After only about six months at Zurich, Einstein began to get feelers from the German University at Prague, which anticipated the opening of a full professorship at double his current salary, and with less teaching responsibilities. The offer sounded tempting, and negotiations continued. Meanwhile, Mileva gave birth to their second son, Eduard, aka Tete, in July 1910. Tete was a fussy baby. This served to increase Mileva's depression rather than alleviate it. In January 1911, Einstein was offered the faculty position in Prague, and he accepted it. The Einstein family arrived there in April. Prague was hot and buggy, the water was brown, the air was sooty, and the Germans (10 percent of the population) looked down on the Czechs (who were the majority at 90 percent). Further, about half of the Germans were Jewish, so the other half didn't mix well with them. Mileva was extremely unhappy, to say the least. Einstein was treated like a celebrity at the university, but his research was going badly. To extend his theory of special relativity, he was going to have to abandon normal Euclidean geometry for more exotic possibilities. In addition, the quantum was perplexing him, and he didn't have anybody to bounce ideas off, as he had with Besso.

MEETING MARIE CURIE

In the fall, the first Solvay Conference was held in Brussels. Named for and instituted by Ernest Solvay, a wealthy chemical industrialist, these conferences brought together the best scientific minds so they could update each other and share ideas. Einstein met Marie Curie, who was rumored to be having an affair with French physicist Paul Langevin. Asked for a comment about this rumor by an eager press, Einstein said, "Madame Curie is a simple, honest woman, almost buried under her duties and obligations. She has sparkling intelligence but, despite her passionate nature, is not attractive enough to be a danger to anyone."[1] How's that for a left-handed compliment?

RETURNING TO ACADEMIC HOME BASE

Prior to the conference, Einstein had taught a short course at Zurich. Meeting up with his old friend Marcel Grossmann, now a dean at the ETH, Einstein was asked if would be interested in a faculty position there. Einstein said yes indeed; he knew Mileva would have responded even more positively. Heinrich Zangger, now a dean of forensic medicine at Zurich, pressed the ETH administration even further to favor Einstein with an appointment. After the usual snags, an offer was made in February 1912, and Einstein accepted it. Before moving to Zurich, Einstein traveled to Berlin, where he visited his cousin Elsa (see chapter 9). Elsa was divorced in 1908 from Max Löwenthal, with whom she had two daughters, Margot and Ilse. The cousins reminisced about the old days and enjoyed each other's company tremendously.

In the summer of 1912, the Einsteins moved to Zurich, delighted to escape from Prague. One of the first things Einstein did was to call on Grossmann. He recalled that Grossmann's thesis dealt with more exotic non-Euclidean geometry, and Einstein had decided that in generalizing relativity he also needed more generalized geometry, but he was without a clue in that department. "Grossmann, you've got to help me," he implored.[2] Grossmann was eager to assist, but only with the mathematics; he wanted no part of the physics. Einstein was quite amenable to that arrangement. They wrote two papers together, with the mathematics and the physics clearly separated. The equations they came up with still didn't achieve what Einstein wanted: a simple mathematical formulation that related gravity to the fundamental curvature of space-time that was general enough that it didn't depend on whatever coordinate system the equations were formulated in.

This qualifies as save number three for Grossmann (see chapter 9 for the first two), because his help was just what Einstein needed. The reason the equations didn't work was because Einstein made an error when he checked them and didn't catch it at the time. In Einstein's defense, the calculations involved extremely complicated mathematical forms, and so it was easy to make a mistake. On the other hand, all those skipped classes at the ETH may have taken their toll. In Einstein's words, "Never before in my life have I troubled myself over anything so much, and that I have gained great respect for mathematics, whose more subtle parts I considered until now, in my ignorance, as pure luxury! Compared with this problem, the original theory of relativity is childish."[3]

THE BIG TIME

While the general relativity equations simmered, the rest of life continued at its merry pace. The Einsteins traveled to Paris, where they stayed with the Curies. The families became friendly, so the Einsteins planned to return in the summer for some hiking. Teaching and research continued at their normal pace, but then a big academic break came in the form of visitors from Berlin.

Max Planck and physicist Walther Nernst arrived with a directive from Fritz Haber, the new director of the Kaiser Wilhelm Institute of Physical and Electrochemistry: get Einstein. The offer they made was overwhelming. Einstein would be a full professor at the University of Berlin as well as the director of the Kaiser Wilhelm Institute of Theoretical Physics (when it was built). His salary would be the maximum allowed for any professor at the university, and he would have zero teaching duties.

The idea of returning to Germany didn't appeal to Einstein. He knew Mileva would hate it there. The directorship didn't interest him, but the lack of teaching duties did. He was beginning to feel worn down at the ETH, and he suspected he might be very close to finishing general relativity, his all-consuming interest at this time. His friend from undergraduate days, Louis Kollros, recalls Einstein saying, "The Berliner gentlemen speculate that I am an award-winning Chicken-hen, but I do not know if I can still lay eggs!"[4] Einstein soon accepted but delayed the start time until the spring of 1914. This allowed him to attend conferences, travel, hike with the Curies, struggle with the general relativity equations, and arrange a secret meeting with Elsa. It also represented the calm before the storm.

TROUBLE IN BERLIN

The impending move to Berlin made Mileva even gloomier than ever. She didn't much care for Germans. In addition, Einstein's relatives would be in proximity, in particular his mother and that pesky cousin Elsa. In late December, Einstein sent Mileva to Berlin to find an apartment, since he had little interest in housing. She returned to Zurich even more unhappy and even a little suspicious of Elsa, who offered to help find housing close to her.

When moving time came in late March, Tete developed multiple illnesses, which led to his exhaustion. Doctors recommended recuperation at a spa, so Mileva and the boys left for Locarno, a resort town in the Italian colony of Switzerland. This left Einstein free to move to Berlin by himself. He canceled a physics meeting to arrive early and spend time with Elsa. At the university, Einstein was charting a somewhat different course than his sponsors—Haber, Nernst, and Planck—must have expected. Einstein wanted experimental tests of the general relativity theory, not a research program in quantum theory.

It was late April by the time Mileva and the boys arrived in Berlin. Einstein was uncommunicative at home, and Mileva felt like she was being ignored. Even Hans Albert complained that his father had become "nasty" since the move to Berlin. Mileva suspected Einstein was spending time with Elsa. They had a giant argument in July. As a result, Einstein moved out of the apartment to stay with his Uncle Jakob, and Mileva and the boys began living with Clara Haber. Their major form of communication became notes passed through Fritz Haber, a role Haber neither expected nor appreciated. Finally, a separation agreement was negotiated, with Mileva and the boys moving back to Zurich, and visitation of the boys taking place only on neutral ground, and never at Elsa's. Besso came to Berlin to accompany Mileva and the boys back to Zurich, and Fritz Haber went to the train station with Einstein to see the boys off.

EVEN BIGGER TROUBLES

As if the personal war between Einstein and Mileva wasn't bad enough, a far larger conflict broke out at the same time: World War I. With world opinion strongly against Germany, especially after its invasion of Belgium, ninety-three prominent German scholars, politicians, and authors signed a manifesto supporting Germany. One of Elsa's friends, Georg Nicolai, soon circulated a counter-petition urging a stop to the war, pointing out that there would be no victors, only victims. Einstein's pacifism kicked in, and he signed it, but not many others did, and it wasn't widely published.

Einstein found a bachelor apartment near Elsa's, but not too near. The occasional square meal and delightful company were great, but her mothering instincts became a bit much at times. Much as he missed the boys, he didn't miss Mileva, and the quiet allowed him to plunge into his work full bore. He pursued general relativity so relentlessly that he cared little about his own health. He was living the way he had described to Elsa in a letter from several months earlier, "I have firmly decided to bite into the grass with a minimum of medical help when my little hour has come. But until then, I plan to sin away as my wicked soul bids me. Diet, smoking like a chimney, working like a horse."[5] Unfortunately, the horse was pulling a faulty load. When Einstein and Grossmann worked together more than two years earlier, they'd had to choose between two fundamentally different mathematical techniques to formulate their equations. Their first choice was eliminated by Einstein on the basis of a lengthy calculation, in which he had made an error. Thus, he labored away using the other tensor, which required many adjustments, but he still couldn't make it fit. Besides, he was no longer in close touch with Grossmann, who was still in Zurich at the ETH. Fortunately, Einstein piqued the curiosity of Germany's greatest mathematician, David Hilbert.

David Hilbert (1862–1943). From Wikimedia Commons, user Mschlindwein.

In June 1915, Einstein traveled to Göttingen to present a series of talks and had a very pleasant time with David Hilbert. Unlike Marcel Grossmann, who harbored a certain distaste for physics, Hilbert liked it and had been following Einstein's career with interest. Besides, Hilbert was one of the few besides Einstein who had signed Nicolai's antiwar petition. When Einstein outlined his struggles to generalize relativity, Hilbert, who was familiar with the tensors, instantly understood what Einstein was trying to accomplish. One of Hilbert's colleagues, Felix Klein, wondered about Einstein's ability to carry out his ambitious plans. "Einstein is not innately a mathematician, but works rather under the influence of obscure physical-philosophical impulses."[6]

Einstein spent a restless summer with a nagging feeling of something amiss in his analysis. Hilbert spent some of his summer vacation on Einstein's equations and also realized something must be wrong. Hilbert took the additional step of encouraging one of his brightest assistants, Emmy Noether, to work on a generalization of similar equations, and she formulated and proved Noether's first theorem. This very general theorem shows that for every symmetry of a physical system, there is a corresponding conservation law. In a very elegant way, she demonstrated that Einstein's general relativity automatically included energy and momentum conservation. Before her work, there was some concern about whether energy would be conserved, and it could only be answered by lengthy, laborious calculations. Einstein wrote to Hilbert: "Yesterday I received from Miss Noether a very interesting paper on invariants. I'm impressed that such things can be understood in such a general way. The old guard at Göttingen should take some lessons from Miss Noether! She seems to know her stuff."[7]

ERRORS DISCOVERED

Then, Einstein went back to work and discovered his calculation error from almost three years earlier that seemed to make the second formulation preferable to the first one. Next, he repeated the calculation that knocked out the first choice and found another error. Finally, he was able to return to the first choice that he and Grossmann had made. With great effort, he reformulated his equations to present to the Prussian Academy of Sciences. Acknowledging the giant difference from the equations he presented to the same group the prior year, he said, "No one who has really grasped it can escape the magic of this theory."[8] Writing to his friend Paul Ehrenfest about what academy members must have thought, Einstein said, "That fellow Einstein suits his convenience. Every year, he retracts what he wrote the year before."[9]

Having equations is a good start, but next must come solutions to these equations. Einstein was able to show that his equations predicted a small deviation in the orbit of Mercury. That small effect matched experimental results almost exactly. More support was needed, but this would take a while. Meanwhile, Hilbert published his own equations, which were quite similar. This caused a minor rift, and Hilbert claimed to understand Einstein's work better than Einstein himself. This may have been true, but Hilbert's paper acknowledged Einstein's priority, so the difficulty went no further.

Einstein had scored a major coup on the old Newtonian theory. Now, time and space not only depended on the observer; they were deformed by the presence of mass. Einstein's general theory of relativity says that throughout the entire universe, the presence of mass curves space-time. A two-dimensional analogy is provided by a stretched blanket: A small marble will roll straight across an empty blanket, but if a bowling ball is placed on the blanket, the marble's path will be curved by the presence of the ball.

PERSONAL STRUGGLE

Einstein was elated that his difficulties with formulating a general theory of relativity were finally over, and now he made a stab at resolving his difficulty with Mileva. Separation didn't seem to be a long-term solution, so he asked her for a divorce. Her response was to deny him access to the boys. She then proceeded to have a nervous breakdown. Einstein's friends in Zurich—Besso and Zangger—helped with the boys, but Einstein wasn't doing very well health-wise himself. What with all his nonstop working, his family struggles, the ongoing war, and the rising tide of anti-Semitism in Berlin, he developed horrible stomach pains and thought he was going to die. After a period of denial, followed by doctor visits and more denial, he came under Elsa's care by moving to an apartment next door to hers.

SOLUTIONS

Albert Einstein, far left, and Willem de Sitter (1872–1934), far right, about 1923. Photographed by H. van Batenburg, 1923. Leiden Archives, from Wikimedia Commons, user Mdd.

Meanwhile, Einstein's equations were being studied by astronomers. Karl Schwarzschild found solutions both inside and outside massive objects such as stars, and Willem de Sitter applied the equations to the universe as a whole.

Even though Einstein battled stomach and family problems, he and de Sitter carried on a two-year running debate about the nature of the entire universe in relationship to the field equations of general relativity. Others, especially the mathematicians from Göttingen, joined the debate, and papers and presentations flew back and forth with details, arguments, and counterarguments, all quite civil. The biggest problem turned out to be the status of the universe in terms of expansion, contraction, or remaining static. Einstein finally added a term to his equations in order to keep the universe static, because de Sitter convinced him that the fundamental equations implied expansion. Einstein called the term the *cosmological constant*, and its addition annoyed him because it marred the equation's simplicity (more on this constant in chapter 11).

SOCIETAL STRUGGLES

In scientific circles, Einstein's reputation and notoriety grew, but the war droned on. Jewish refugees from the Eastern front flooded into Germany, unleashing a tide of anti-Semitism. Since Einstein was so prominently antiwar, his support for refugees marked him as vaguely anti-German, and because he was Jewish, his scientific standing among right-wingers was tarnished. At first, he was too busy to notice much other than his work, but the situation became hard to ignore. The war ended in November 1918, but anti-Semitism was just barely hitting its stride.

ZIONISM

Early in 1919, with perfect timing, Kurt Blumenfeld showed up at Einstein's door. He wanted to speak to Einstein about the Zionist cause. Einstein had been exposed to Zionism before and had not displayed much interest, but he was especially vulnerable at this time.

Kurt Blumenfeld (1884–1963). bpk, Berlin / Central Zionist Archives, Jerusalem / Art Resource, NY.

Stomach problems, the completion of his general relativity project, his pending divorce, the plight of war refugees, the end of the war, the rising tide of anti-Semitism, and the establishment of Palestine as a Jewish state by the Balfour Declaration all conspired to make Einstein receptive to Blumenfeld's ideas. Blumenfeld quotes Einstein's reaction: "I am against nationalism but in favor of Zionism. The reason has become clear to me today. When a man has both arms and he is always saying I have a right arm, then he is a chauvinist. However, when the right arm is missing, then he must do something to make up for the missing limb. Therefore, I am, as a human being, an opponent of nationalism. But as a Jew I am from today a supporter of the Jewish Zionist efforts."[10] Einstein's change of heart that day led to several later consequences, as we will see.

RESOLUTION

The world war ended only a few months before the personal one between Einstein and Mileva did. For more than a year, negotiations between the two dragged on by letter. The process was interrupted by illnesses, hospitalizations, hyper-inflation in German currency, even a joint offer of employment from the University of Zurich and the ETH—both of which were the work of Zangger. At long last, the divorce agreement was sealed. The final terms included a deposit of forty thousand marks in a Swiss bank from which Mileva could draw the interest but not the principal, and the proceeds from the Nobel Prize if Einstein won it.

The source of the forty thousand marks is unknown, and the Nobel Prize proceeds require a little more explanation. It might seem presumptuous on Einstein's part, but he had been nominated for the Nobel Prize eight times since 1910, and his recent completion of the general relativity theory seemed like it would make him a shoo-in. Actually, the story of why Einstein didn't win the Nobel Prize earlier is rather lengthy and quite political, involving prejudices against Jews, pacifists, as well as judging densely mathematical theoretical physics in comparison with easier-to-understand experiments.

With the divorce finalized in February, Einstein, currently living with Elsa and her two daughters, Margot and Ilse, was free to marry. But then a curious thing happened. An odd discussion took place. The question was first suggested by Ilse's boyfriend in a half-joking way, should Einstein marry Elsa or Ilse? The matter was discussed, and Einstein graciously said he would be agreeable to marry either one, and they should be the ones to choose. After unrecorded discussions between mother and daughter, Einstein and Elsa were married in June. They had separate bedrooms at opposite ends of the apartment because Einstein "snored too much."

EXPERIMENTAL SUPPORT

Much of the rest of 1919 was spent in eager anticipation of solar events. The next eclipse of the sun, which would occur May 29, provided an opportunity to test general relativity. According to Einstein's theory, stars sending out light rays that traveled close to our sun would be bent by its mass, and the star's apparent direction would shift. The predicted effect was quite small. It could only be measured by high-quality astronomical equipment, only a small swath of Earth's landmass would be able to see the eclipse, and bad weather could negate the whole effort. The world's astronomers relished the chance to have an impact on Einstein's new theory, many hoping to disprove it. Only one country was in the position of being able to fund an expedition that would get the right equipment to the right place at the right time: England.

Arthur Eddington was just the right fellow to perform this experiment. His skills were a rare combi-

Sir Arthur Eddington (1882–1944).
From Wikimedia Commons, user Mu.

nation: high mathematics and practical astronomy. Willem de Sitter had sent him a copy of Einstein's work, and he grasped it immediately. Eddington explained general relativity to his colleagues in England, although ideas from Germans were not easily accepted in those war years. The story goes that he was congratulated on being one of three people who understood the theory. Eddington hesitated and was accused of false modesty. He then said, "On the contrary, I was trying to imagine who the third person was."[11]

The eclipse's path of totality ranged across the south Atlantic from South America to Africa, so Eddington and his assistant E. Cottingham were dispatched to the island of Principe, off the west coast of Africa. Andrew Crommelin and Charles Davidson were sent to Sobral, Brazil, in case weather spoiled the data from Eddington's efforts. The Newtonian conception of gravity predicted a shift of a star's position of 0.86 second of arc, and Einstein's general relativity predicted a 1.74-second shift. The anticipated eclipse finally occurred, the telescopes (four-inch-diameter models) were ready, and the clouds parted reluctantly to yield some data. It took some time to return to England and analyze the data, but the final result was 1.79 seconds. Close enough, within experimental uncertainty, to support Einstein's general theory of relativity.

The London *Times* headline read: "Revolution in Science—New Theory of the Universe—Newton's Ideas Overthrown." The story was picked up in the United States, and before long, Einstein became a worldwide celebrity. He joked to reporters, "Today in Germany, I am called a German man of science and in England I am represented as a Swiss Jew. If I come to be regarded as a *bête noir*, the description will be reversed, and I shall become a Swiss Jew for the Germans and a German man of science for the English."[12]

MORE ANTI-SEMITISM

He didn't have long to wait for the latter to occur. After the war, the German mark dropped to an all-time low, and unemployment was rampant. A scapegoat was found close at hand: Jews—and Einstein represented a prominent example. A group calling itself German Scientists for the Preservation of

Pure Science organized a rally at which relativity was dismissed as a "publicity stunt," and Einstein called a plagiarist. Oddly, Einstein attended the rally and later wrote a letter to the editor of *Berliner Tageblatt*, a liberal German newspaper, saying that the criticisms were superficial and made little sense. Further, he pointed out that one of the rally's behind-the-scenes organizers, the Nobel Laureate Phillip Lenard, had never achieved anything in theoretical physics. Later, Einstein regretted his loss of temper, saying, "Everyone is entitled to one act of stupidity."[13] Ironically, Mileva had attended Lenard's lectures in Heidelberg in 1897 and had written to Einstein about his experiments with the photoelectric effect. Further, one of Einstein's 1905 papers explained the photoelectric effect in terms of the quantum. To top all, as we'll see shortly, Einstein's Nobel Prize was awarded for . . . you guessed it, the photoelectric effect. Lenard's anti-Semitic views became even stronger. He created more trouble for Einstein with his book *One Hundred Authors against Einstein*. Eventually, Lenard became a Nazi with direct access to Hitler.

ANOTHER FRIEND

In 1920, Einstein participated in a colloquium along with German physicist Max von Laue and Max Planck at the University of Berlin's Institute of Physics. One of the younger colloquium members was Leo Szilárd.

Although Leo Szilárd was almost twenty years younger than Einstein, the two of them became friends, walking home together after colloquia, with Szilárd often stopping over for tea. After hearing that a German family had perished when

Leo Szilárd (1898–1964). From Wikimedia Commons, user Panoptik~commonswiki.

a refrigerator leaked toxic chemicals in their apartment, the two of them brainstormed a refrigerator that had no moving parts and eventually patented it. A working model of their refrigerator was actually built, but it wailed like a banshee and was never a practical success. As we'll see in chapter 12, Szilárd played a strong role in Einstein's activities in World War II.

TRAVELS

Early in 1921, Einstein heard from Kurt Blumenfeld again, this time with an invitation. Blumenfeld carried a telegram from World Zionist Organization president Chaim Weizmann.

Weizmann was a biochemist who emigrated from Russia to England and had deep commitments to Zionism. He invited Einstein to join him on a trip to America to raise funds for settling Palestine, and especially for the Hebrew University of Jerusalem. To their surprise, Einstein accepted, and

Chaim Weizmann (1874–1952). From Wikimedia Commons, user Jappalang.

he and Elsa set sail for America in March 1921. During the trip, Einstein tried to explain the theory of relativity to Weizmann. Upon arrival in New York, Weizmann said, "Einstein explained his theory to me every day, and by the time we arrived I was fully convinced that he really understands it."[14]

THE NOBEL PRIZE, FINALLY

The Nobel committee was hamstrung by indecision in 1921, thanks partly to Philip Lenard, who opposed Einstein at every turn. The following year, however, changes were made to the committee membership, and Einstein was awarded the prize, officially stated as 1921's prize. The citation read, "Services to theoretical physics, and especially for his discovery of the law of the photoelectric effect." When the prize was announced, Einstein and Elsa were on a trip to the Far East, which included Palestine. Besides the honor of the prize, there was an extremely practical side to this award. The money, about ten times an average professor's annual salary, went to Mileva, even though the divorce agreement said she only got the interest. (Einstein hid this fact for a long time.) She promptly bought three apartment houses in Zurich and lived on the top floor of one of them.

A QUANTUM FRIEND

Einstein was one of the first to take Planck's quantum as a valid representation of reality, even though Planck himself had regarded it as a mathematical convenience. He was equally quick to embrace Niels Bohr's use of the quantum to explain the radiation of light from the hydrogen atom in 1913. Quantum mechanics, however, had taken some unsatisfactory turns in Einstein's view, partly due to Danish physicist Niels Bohr.

Einstein's difficulty with quantum mechanics was rooted in cause and effect. Bohr, Werner Heisenberg, Wolfgang Pauli, Erwin Schrödinger, and Paul Dirac had developed equations that injected an element of uncertainty into the fundamental description of reality, typified by Heisenberg's uncertainty principle: It is impossible to measure exactly the position and momentum of any particle at the same time. Although this principle flows directly from the dualistic wave/particle nature of particles, it implies that probabilities are involved at a very fun-

Niels Bohr (1885–1962). From Wikimedia Commons, user Craigboy.

damental level. Einstein couldn't buy this. Although this is a significant philosophical idea, a way to express his view is a quote from him: "I can't conceive of God playing dice."[15]

Niels Bohr provided the counterpoint to Einstein's argument by saying, "Einstein, stop telling God what to do with his dice."[16] The serious parts to the Einstein/Bohr debates took place over many years, and in many different public and private forums, starting with the 1927 Solvay Conference. The discussions often followed this format: Einstein comes up with an argument to cast doubt on the quantum mechanical theory; Bohr thinks about it for a while, then demolishes the argument. Supposedly, one such debate took place on a streetcar. They were so engaged that they missed their stop and went to the end of the line. They got back on the car, started arguing again, then missed the stop again going the other way.

Bonus Material: Bohr/Einstein Internet interview. See To Dig Deeper for details.

The debates between Einstein and Bohr were quite civil, and the two had the greatest mutual respect and were quite good friends. In a letter to Bohr, Einstein said, "Not often in life has a human being caused me such joy by his mere presence as you did."[17] Bohr became involved in World War II, and will appear again in chapter 12.

Niels Bohr and Albert Einstein, about 1925.
From Wikimedia Commons, user Craigboy.

MORE FAMILY STUFF

Besides the first Einstein/Bohr debate at the Solvay Conference, 1927 saw a signal family event for the Einsteins: Hans Albert graduated from the ETH with a degree in civil engineering. But then the other shoe dropped. Hans Albert announced he was going to marry Freida Knecht. Einstein hit the roof. Freida was nine years older than Hans Albert and was only four feet eleven inches tall. Ironically, Einstein reacted in the same way his mother had about Mileva, and Hans Albert reacted like his father had—stubbornly. The marriage took place on May 7, 1927; the children of the marriage were not dwarfs; and Einstein later admitted, reluctantly, that Hans Albert's marriage was happier than either of his.

UNIFIED FIELD BEGINNINGS

It wasn't long after general relativity that Einstein began thinking about his next project. His goal was to unify gravity and electromagnetism into what he called a unified field. But action followed thought slowly in this case. Family events, the experimental support of general relativity, his new-found celebrity status, his travels, and even some health issues delayed his work substantially. Further, he was not alone in his pursuit of unification. Other theorists proposed ideas, none of which reached the level Einstein sought. Finally, in 1929, Einstein pursued a variation on a theory proposed by Hermann Weyl and decided he was on the right track.

As he prepared his paper, word somehow leaked out, and reporters gathered outside his house. The paper was published by the *New York Times*, dense equations and all. Within months, Einstein abandoned that particular theory on mathematical grounds, but he pursued this holy grail of the unified field theory the rest of his life. He never stopped thinking about it, but he recognized the frustration, saying, "Most of my intellectual offspring end up very young in the graveyard of disappointed hopes."[18]

The year 1929 also marked a significant birthday for Einstein: he turned fifty. Several of his friends went in together and bought him a twenty-three-foot sailboat, which he called *Tümmler*, German for "porpoise." He and Elsa then sank much of their life savings into building a cottage on a lake at Caputh, near Potsdam, only a short drive from Berlin. It was quiet and peaceful at Caputh, and the Einsteins loved it as an escape from the ever-increasing difficulties in Berlin.

The Einstein lake cottage at Caputh. From Wikimedia Commons, user Oursana.

WOMEN FRIENDS

Perhaps the relaxation at the lake house was a little more than Elsa bar-
gained for. Several wealthy widows came for visits in their chauffeur-
driven cars, and Einstein would go away with them to concerts and perhaps
stay the night. Elsa saw many women throw themselves at Einstein. She
was a good sport about it for the most part. But when Einstein asked Elsa,
who controlled finances, for spending money for incidentals on one of
these outings, Elsa balked and refused to give him money for his "hussy."
Einstein lost his temper and took the "hussy" sailing. The next day, Elsa
became even more angry when she found a skimpy bathing suit on the
boat. As a live-in maid put it, "Einstein loved beautiful women, and they
loved him in return."[19]

MORE TRAVEL

In a visit to Caltech in 1931, Einstein was delighted to meet Edwin Hubble (see chapter 12). Hubble had telescopic evidence that the universe was expanding, which allowed Einstein to remove the cosmological constant from his general relativity equations. In California, Einstein had lunch with the Hollywood actor Charlie Chaplin and attended the premier of his movie *City Lights* with him. Einstein later said, "Just as in his films, Chaplin is an enchanting person."[20]

Travels, visiting professorships, and extended weekends at Caputh were becoming more commonplace as the anti-Semitism in Berlin increased. Finally, in 1932, Einstein could see the handwriting on the wall. He accepted an offer from Abraham Flexner to conduct pure research at the Institute for Advanced Study in Princeton, New Jersey, but he anticipated only half-time there and half-time in Berlin. Meanwhile, German anti-Semitism was stronger than ever, and Einstein became more politically active, speaking out strongly against the Nazis. Before they left for a speaking tour of the United States in December 1932, Einstein said to Elsa, "Before you leave our villa this time have a close look at it." "Why?" asked Elsa. Einstein answered: "You'll never see it again."[21]

HITLER

Sure enough, while the Einsteins were in America, Adolf Hitler became chancellor, and Einstein's apartment in Berlin and the lake house were searched and his beloved sailboat confiscated. With some assistance from Philip Lenard and other prominent anti-Semites, Einstein had become number one on Hitler's hit list. There was a price on his head in Germany. One of the milder things Einstein said about Hitler was "The world needs heroes and it's better they be harmless men like me than villains like Hitler."[22]

BAD NEWS

Since the Institute for Advanced Study job didn't begin until October, the Einsteins left the United States for Belgium, staying with the royal family under guard. Much to his dismay, Einstein learned that Tete had had a nervous breakdown, and so he traveled to Zurich at great personal danger. Tete told his father he hated him, and that he had "cast a shadow over his life."[23] Heinrich Zangger thought Tete might be able to recover enough to live a normal life, but that didn't occur. Tete's schizophrenia kept him in and out of sanitariums for the rest of his life. Einstein left Zurich depressed, never to see Tete or Mileva again.

Einstein traveled to Oxford and gave several lectures. During his stay there, he learned that his friend Paul Ehrenfest had shot his mentally deranged son and then committed suicide. With heavy heart, the couple sailed for America and had barely gotten settled in Princeton before more bad news arrived. Elsa's daughter Ilse was gravely ill with tuberculosis in Paris. Ilse died soon after her mother arrived, and Elsa returned to Princeton with Ilse's ashes, heartbroken. The bad news kept coming. The following year, Einstein got word that his friend Marcel Grossmann had died of multiple sclerosis. Immediately, Einstein wrote to his widow: "I remember our student days. He the irreproachable student, I myself disorderly and a dreamer. He on good terms with the teachers and understanding everything, I a pariah, discontent and little loved. But we were good friends."[24] Even closer to home, Elsa developed an eye problem. Doctors told Einstein her condition was far worse. She had heart and liver problems that wouldn't give her very long to live. She must have wondered at Einstein's attentiveness, but that didn't last long. Elsa died in 1936.

Einstein kept plugging away at his unified field theory with little success, although he had some collaborators at Princeton. His rejection of quantum mechanics played some part in his lack of progress. He acknowledged this by saying, "I must seem like an ostrich who forever buries his head in the relativistic sand in order not to face the evil quanta."[25] He also abandoned the physical insights he had relied on in his youth and became more dependent on formal mathematical arguments, never his strongest suit.

MAJA

In 1939, the war in Europe again influenced Einstein's life. Mussolini moved to copycat Hitler's anti-Semitic laws, which put Maja and Paul Winteler at risk in Italy. The photo shows Maja and Paul in happier days with the rest of the Winteler family.

Maja Einstein (1881–1951) second from left, Paul Winteler (1882–1952) third from left, and the rest of the Winteler family, about 1900. From Wikimedia Commons, user Freigut.

They had lived in a pleasant villa near Florence and were patrons of the arts. Besides his law practice, Paul fancied himself a painter. Maja's piano talents were well-known, especially playing on the Blüthner piano Einstein had given her. They decided they had to leave Italy and reunite after the war. Einstein offered to host them at Princeton, but Paul's health wasn't good enough to make the trip. Paul went to Zurich to live with his sister Anna and her husband Michele Besso, Einstein's good friend. Maja came to Princeton to live with her brother. In later years, Maja and Albert had come to resemble each other to the point that, from the back, it was hard to tell them apart.

MORE FAMILY MATTERS

Not long after Einstein arrived in the United States, he applied for citizenship, realizing he could keep his Swiss credentials also. Citizenship was granted several years later, in 1940. Hans Albert left Switzerland in 1938, and settled in South Carolina, where he worked for the US Department of Agriculture studying sediment transport. The relationship between father and son was a complicated one, but improved as the years went on. Hans Albert had one son who survived to adulthood, Bernard, and an adopted daughter, Evelyn, born in Chicago in 1941.

Bernard eventually studied at the ETH and worked for defense contractors in the United States and Switzerland. He married Aude Ascher and they had five children: Thomas and Eduard, who live in California; Paul in France; Mira in Israel; and Charlie, who lives in Switzerland.

Bernard Einstein (1930–2008), about 2003. Courtesy of Thomaseinstein, from Wikimedia Commons.

STILL MORE WOMEN

Even in America, Einstein still had affairs with women. One particularly interesting one was Margarita Konenkova.

Margarita was a lawyer. She spoke five languages and was very intelligent. She was in Princeton in 1942, accompanying her artist husband, Sergei, who had been commissioned to sculpt a bust of Einstein. When Margarita and Sergei arrived at Einstein's home, there was apparently instant attraction between Margarita and Einstein, and she invited him to a weekend party at a friend's summer home on Long Island. Einstein accepted, and an interesting relationship began, which was discovered by the public only in the late 1990s when a Konenkova relative ran across some letters and pictures from the 1945–1946 era. Of particular interest was that Margarita had been listed as a spy in the memoirs of a Soviet spymaster. Could she have been picking Einstein's brain for details about the Manhattan Project? If so, these would have

Einstein and Margarita Konenkova (1896–1980). © Sergey Konenkov/Sygma/Corbis.

been slim pickings. As we'll see in chapter 12, Einstein's involvement with the A-bomb was only before the beginning of the project, and through his friend Leo Szilárd. Einstein did contribute to the war effort, but in a different way. He was asked if he would submit his original papers on relativity for auction. Since they had long ago been discarded, he offered to rewrite them. He did so, along with another paper. They fetched $11 million for the war bond campaign.

RACISM

After the war ended, Einstein's health wasn't very good. He seldom accepted honorary degrees, saying they were "ostentatious." Yet, in 1946, Einstein traveled the sixty miles from Princeton to Lincoln University, near Oxford,

Einstein at Lincoln University, 1946. Photographed by Peace Photo. From the Shelby White and Leon Levy Archives Center, Institute for Advanced Study, Princeton, NJ, USA.

Pennsylvania. Lincoln was the first degree-granting historically black university. He stood outdoors for hours, as there was no building on campus large enough to accommodate the crowd that showed up to hear him. In his address, he said, "There is separation of colored people from white people in the United States. That separation is not a disease of colored people. It is a disease of white people. And, I do not intend to be quiet about it." This was only the tip of the iceberg of Einstein's civil rights activities.

It all started in 1931, when W. E. B. Du Bois wrote a letter to Einstein in Berlin, asking if he could write a small piece for the NAACP's journal *The Crisis*. Einstein did respond. Here's an excerpt: "It seems to be a universal fact that minorities, especially when their Individuals are recognizable because of physical differences, are treated by majorities among whom they live as an inferior class. . . . The determined effort of the American Negroes in this direction deserves every recognition and assistance."

Subsequently, Einstein met and became friends with Paul Robeson in 1935 when Robeson came to perform on campus at Princeton. Robeson, born in Princeton, was the valedictorian at Rutgers University, became an All-American football player, got a law degree from Columbia, and played in the NFL briefly. He was a good singer and actor, and played in Broadway plays and Hollywood films such as "Emperor Jones," the first film to star an African American. Robeson grew to be disenchanted with treatment of black Americans, and became sympathetic with communism. Subsequently, he was blacklisted in Hollywood and his passport was revoked. Nevertheless, Einstein invited Robeson to come for a visit in 1952, along with his friend Lloyd Brown. When Robeson briefly left the room, Brown told Einstein it was "an honor to meet a great man." Einstein fired back, "You came here with a great man."

Other highlights of Einstein's antiracism include his friendship with opera singer Marian Anderson. She performed on Princeton's campus in 1937 to rave reviews, but was turned away from the Nassau Inn because of her color. Einstein invited her to stay at his home, and she stayed there every time she traveled to Princeton. Einstein became a member of the NAACP in 1942 and was often seen strolling the streets in the black community, handing out candy to children.

ZIONISM AGAIN

Einstein's friend Chaim Weizmann had become the first president of Israel. When Weizmann died in 1952, Israel's prime minister, David Ben Gurion called Einstein "the greatest Jew on earth," and offered him Israel's presidency. Einstein's answer was:

"I am deeply moved by the offer from our State of Israel (to serve as President), and at once saddened and ashamed that I cannot accept it. All my life I have dealt with objective matters, hence I lack both the natural aptitude and the experience to deal properly with people and to exercise official functions." This answered the question in just the way Ben-Gurion hoped for. Before he received Einstein's answer, Ben-Gurion said to an

Albert Einstein and David ben Gurion (1886–1973).
Used with permission from Science Source.

aide: "Tell me what to do if he says yes! I've had to offer him the post because it was impossible not to, but if he accepts we are in for trouble!"

MATHEMATICS—AGAIN

From 1942 on, Einstein had company of the highest order at the Institute for Advanced Study—mathematician Kurt Gödel.

Einstein and Gödel had met as early as 1931, but many things had occurred since then. When Gödel succumbed to Abraham Flexner's charms and decided to come to Princeton permanently, Einstein was delighted. The two men had large divergences in personality and opinion about almost everything, but there was a strong bond of mutual respect. Although Einstein might have found it more convenient to work from his home at 115 Mercer Street, he dutifully walked to and from the institute's office building every day. A colleague named Oskar Morgenstern said that Einstein told him he went to the office "just to have the privilege of walking home with Kurt Gödel."

For Einstein, it must have brought back memories of walking home from the Bern patent office with Michele Besso. At Princeton, Einstein and Gödel were beginning to feel like museum pieces. Most of their colleagues would agree with their views out of deference or disagree silently. Gödel was such an introvert that the ideas of others weren't that important. On the other hand, Einstein thrived on interaction. Between the two of

Kurt Gödel (1906–1978) and Albert Einstein at the Institute, about 1950. Photograph by Richard Arens, courtesy AIP Emilio Segre Visual Archive.

them, it was no-holds-barred. In the 1952 US presidential election, Einstein favored Adlai Stevenson, while, according to Einstein, "Gödel has gone extremely crazy. He voted for Eisenhower." Closer to his scientific interests, Gödel came up with one solution to Einstein's equations in which time travel was possible and another in which there was no time at all. Gödel's brilliant mathematics made Einstein wonder. He wrote to Besso in 1954, "I consider it quite possible that physics cannot be based on the field principle, i.e., on continuous structures. In that case, *nothing* remains of my entire castle in the air, gravitation theory included . . ." Clearly, this odd couple thrived on their walks and talks.

For all his adventures, both intellectual and in other parts of his life, Einstein would seem to make a fascinating dinner companion. How could one fail to enjoy a fellow who said, "Whoever undertakes to set himself up as a judge of Truth and Knowledge is shipwrecked by the laughter of the gods."

Albert Einstein. Image courtesy of the Observatories of the Carnegie Institution for Science Collection at the Huntington Library, San Marino, California.

CHAPTER 11

EDWIN HUBBLE AND HARLOW SHAPLEY CLASH/COOPERATE OVER THE UNIVERSE'S SIZE

Now my own suspicion is that the universe is not only queerer than we suppose, but queerer than we can suppose.

J. B. S. Haldane[1]

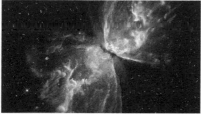

Butterfly nebula. NASA, ESA, and the Hubble SM4 ERO Team.

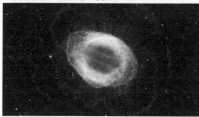

Ring nebula. NASA, ESA, C. R. O'Dell (Vanderbilt University), and D. Thompson (Large Binocular Telescope Observatory).

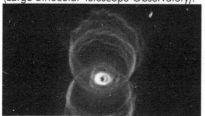

Hourglass nebula. Raghvendra Sahai and John Trauger (JPL), the WFPC2 science team, and NASA.

Since this book is about humans, what human is responsible for these pictures?

EDWIN POWELL HUBBLE

Technically, Hubble didn't take these pictures himself; they are from the Hubble Space Telescope, which was named after him. The next time you're surfing the Internet, you might want to visit Hubble's website for the full-color originals and other lovely color photos that boggle the mind.[2] Truly, looking at these images might make you feel puny in comparison to the vast size and complexity of the universe, but it might also make you feel pleased that we human beings have learned a lot about this giant place in which we live.

Since this wonderful telescope bears Hubble's name, it's pretty easy to infer that he must have been the right fellow, at the right place, at the right time, armed with the right instruments. Right.

So, what did Hubble do that led to this giant leap in our understanding of the universe? Let's cut to the chase. In 1929, he wrote a paper titled "A Relation between Distance and Radial Velocity among Extra-Galactic Nebulae." The paper, published in the *Proceedings of the National Academy of Sciences*, included this graph:

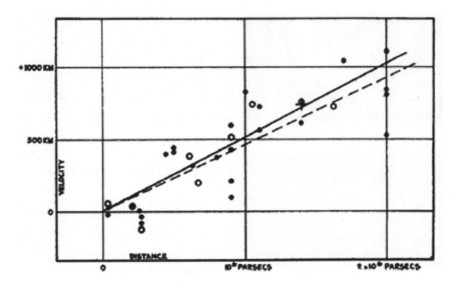

Courtesy of NASA.

That's it. Looks pretty simple, eh? Well, it will take the remainder of this chapter to explain this graph, to introduce the people who contributed to it (both positively and negatively), and to reveal its monster implications for our understanding of the universe.

Let's start with Hubble himself.

EDWIN POWELL HUBBLE (1889–1953)

Virginia Lee James married John Powell Hubble in 1884 in Marshfield, Missouri. Edwin Powell Hubble was the third of their eight children. He seemed to have inherited his mother's good looks and his father's athletic build. As a young child, Hubble had no exceptional qualities, but once he learned to read, he devoured the classics and received high grades in school. The only exceptions were deportment and spelling. Fortunately, deportment grades improved, but spelling remained a sore spot throughout his life. Alas, the mixed blessing of spellcheck came too late for Hubble.

At age ten, he and childhood friend Sam Shelton heard of a lunar eclipse predicted to occur on June 27, 1899. After much pleading from the boys, their parents relented and allowed Sam and Edwin to stay outdoors all night to watch. Sam remembers an unobstructed view and a "magnificent show." Might this have contributed to Hubble's decision to become an astronomer? Or maybe his career choice was related to his appendicitis at age fourteen. He spent several weeks recovering in bed, reading astronomy books.

Hubble's father traveled a lot, but he ruled the roost with an iron fist when he was home. After his law practice failed, John Hubble went into the insurance business. He became a general agent for the Western Department of Greenwich Insurance Company, responsible for some six hundred agents and adjusters in a four-state area. The family moved several times, landing in Wheaton, Illinois, in time for Edwin to attend high school at Central School. A classmate named Albert Colvin, who lived a block away, said Edwin "acted as though he had all the answers, so there was no intimate contact with him."[3]

Academically, Hubble was near the top, but his athletic prowess was even higher. He reached full height during his junior year. He was six feet two inches, a full head taller than all the other boys, save one. At Wheaton Central, he played center on the basketball team. The six-player team was undefeated during the season and won the state championship. In his senior year, he was captain of the track team. In one meet, he won the pole vault, shot put, standing high jump, running high jump, discus, and hammer throw. Between junior and senior year, he worked at a summer job with a surveying crew, and returned ready to become an adult.

At graduation, Superintendent Russell began announcing class honors. "Edwin Hubble, I have watched you for four years and I have never seen you study for ten minutes." He paused for effect. "Here is a scholarship to the University of Chicago."

1903 University of Chicago basketball champs; Edwin Hubble at far left.
Image courtesy of the Observatories of the Carnegie Institution for Science Collection
at the Huntington Library, San Marino, California.

At Chicago, Hubble toed a fine line. His father thought astronomy was "outlandish," so Hubble kept his astronomical studies low-key. The other courses he took were mostly science and math, but they all served as preparation for law school, which was far more pleasing to his father. His athletic activities were curtailed slightly. Hubble loved football and was approached by Amos Alonzo Stagg to play for the University of Chicago. Hubble's father put his foot down, saying there were too many injuries in football. Hubble did some statistical research and pointed out that baseball injuries were just as likely. His father took that as an argument against baseball, so that, too, was banned. Oddly, his father didn't mind boxing, so Edwin Hubble became a well-regarded amateur heavyweight boxer. He also participated in basketball and track, but he played in the shadow of Long John Schommer, a truly gifted athlete who won many All-American honors and led Chicago to two national basketball championships.

The studies went well enough but trailed off a bit toward the end of Edwin's undergraduate days, leaving him with an overall average of B minus. His major distraction was that he had begun to prepare for the Rhodes Scholarship tests. His senior course load included French, Latin, Greek, political economics, and public opinion. He also worked as a lab assistant to physics professor Robert. A. Millikan (winner of the 1923 Nobel Prize in Physics), acted in theatrical productions, and was elected vice president of his senior class. Hubble passed the academic part of the Rhodes Scholarship in good order, and his extracurricular activities and recommendation letters carried the day. He was awarded the 1910 Rhodes Scholarship from Illinois.

In fall 1910, Hubble arrived at Queen's College, Oxford, set to study jurisprudence. In a letter to his sister, he said he had read more than three hundred books over a wide range of subjects, including comparative religion and Russian history. Although it was not discovered until much later, Hubble had spent some time with Herbert Hall Turner, director of the University's Radcliffe Observatory. Contrary to his father's expressed wishes about religion and alcohol, Hubble seldom attended church services and found English ales and French red wines quite interesting. Besides his studies, Hubble also learned the art of rowing, which helped him stay in condition. He also ran track and captained Oxford's baseball team. Holidays were often spent traveling on the Continent, especially Germany, a country that fascinated Hubble. On one such trip, Hubble taught a German naval officer to box, and the officer returned the favor with fencing lessons. Hubble left Germany sporting two small dueling scars, in direct opposition to his father's specific admonition, "In the land and time you will live, the duelist scar is not a badge of honor."[4]

Although the letters from home told Hubble none of it, his father's health was deteriorating. The family moved to Louisville, Kentucky, to be closer to his father's office, but that didn't help. It appeared that he had malaria, but it didn't respond to treatment. Finally, it was discovered that he suffered from Bright's disease (a major kidney ailment) for which there is no known cure. Meanwhile, Hubble had completed his degree requirements early and was contemplating studying literature. When he learned of his father's illness, he immediately requested permission to return home. His father was adamant that he finish, fearing Hubble would never have the financial resources to return to England to complete his studies. Since his law degree was secure, Hubble began a less-demanding study: Spanish. In January 1913, Hubble's father died. Hubble learned of this via cable. He later told his mother that he had sought out a clergyman friend and prayed with him at a chapel.

Hubble finished up at Oxford by becoming president of the Cosmopolitan Club, dining with a foreign countess, dancing into the wee hours, and setting up a party on the river for a close friend. In his last letter to his mother from Oxford, Hubble wrote, "I am glad, awfully glad to feel that I am at last going back to help you just as much as I can."[5]

When Hubble arrived back home, his siblings were amazed to see how different he looked. He wore knickers, a cape, a wristwatch, and carried a cane. His expressions and pronunciation no longer matched theirs, either. The summer passed quickly, with Hubble possibly translating some Spanish legal papers for a Louisville import company. There is no record of Hubble entering legal practice, and the Kentucky bar exam was first administered in 1919. In the fall of 1913, Hubble took a job teaching Spanish, physics, and mathematics at New Albany (Indiana) High School, just a trolley ride across the river from Louisville. He also coached the basketball team to an undefeated season and a third-place finish in the season-ending state tournament. The New Albany High School students enjoyed his knickers, his cape, and his "Oxford mannerisms," and Hubble found that "teaching amused him."[6] Perhaps sensing the snare he was about to fall into, Hubble returned to his first academic love. He wrote to his former astronomy professor, Forest Ray Moulton, inquiring about graduate school and financial assistance. Moulton directed Hubble to Edwin B. Frost at Yerkes Observatory in Wisconsin. Frost was in dire need of a good assistant and promised Hubble a small salary and room and board at Yerkes, beginning in October. In a later letter, Frost suggested Hubble come earlier and meet him at the Evanston campus of Northwestern University for the annual meeting of the American Astronomical Society. Forty-eight papers were read at this meeting's sessions, and only one was greeted by a standing ovation. This paper was so pivotal to Hubble's future work that we must interrupt Hubble's biography here.

The paper that caused such a stir at the 1914 American Astronomical Society meeting was titled "Spectrographic Observations of Nebulae," by V. M. Slipher from the Lowell Observatory in Flagstaff, Arizona. There is a lot of information to unpack here before we can relate it back to Hubble.

NEBULAE

Nebulae is the plural of *nebula*, which is the term used originally to denote any diffuse astronomical object that couldn't be resolved into whatever its constituents might be. An early interest in nebulae came from Charles Messier (1730–1817), who was a comet hunter. As we saw in chapter 2, comets were thought to play a significant role in people's lives, so observers were very interested in comets. Nebulae might be mistaken for comets, so Messier made a list of 110 of them in 1781, so that an astronomer searching for comets would not be distracted. The nebulae were assigned M-numbers by Messier, with the Andromeda Nebula (now known to be the Andromeda Galaxy) called M31.

PERCIVAL LOWELL (1855–1916)

Percival Lowell was the black sheep of the clan, who owned textile factories in Massachusetts. His BA was from Harvard University, in mathematics. At his graduation, he presented a talk on the formation of the solar system. After spending six years in the family business, he began to pursue other interests. First, he fell in love with the Far East. After extensive travels and several books, he turned to astronomy. He then read works by the Italian astronomer Giovanni Schiaparelli, who used the term *canali* to describe formations on Mars. Although Schiaparelli intended the word to indicate channels, which could be natural formations, it was mistranslated as *canals*, implying they were constructed by Martians. Using his vast resources, Lowell selected a site at Flagstaff, Arizona, for an observatory and had installed an Alvan Clark twenty-four-inch refractor telescope and a state-of-the-art John Brashear spectrograph. Lowell intended his observatory at Flagstaff to provide evidence for his Martian civilization theories. Lowell's writings of his theories of civilization on Mars, similar to the works of French astronomer Camille Flammarion, inspired many science fiction writers but ultimately produced no physical evidence.

V. M. SLIPHER

To operate the instruments at Lowell Observatory, Lowell hired V. M. Slipher (1875–1969). Slipher was born in Mulberry, Indiana, located between Lafayette and Kokomo. He earned a BA in astronomy from Indiana University in 1901 and was known as a meticulous, careful observer. Slipher's observations of the Martian surface never produced convincing proof of canals, but he made many other detailed measurements of planets, stars, and

V. M. Slipher (1875–1969). Courtesy of LOBS, from Wikimedia Commons.

nebulae using the telescope and the spectrograph. The paper he presented at the 1914 meeting of the American Astronomical Society was based on his work with the spectrograph, which revealed an extremely unexpected attribute of nebulae—they were moving, some very fast. How did Slipher measure these velocities? They were based on the Doppler effect.

DOPPLER EFFECT

Many of our own experiences contain examples of the Doppler effect. Think about driving along the highway, minding your own business. Suddenly, you hear a dreaded sound behind you and look into your rearview mirror. Sure enough, it's a police car, siren wailing. You glance at your speedometer. Your speed is within the legal limit now, but how fast were you going when you passed that police car about a mile back? Sweat, sweat. Much to your relief, the police cruiser speeds by. But you notice an odd thing. The sound of the siren was higher-pitched when the car was bearing down on you, then lower-pitched when it was driving away.

This isn't your imagination; it's a real phenomenon, called the Doppler effect. When a sound wave is emitted by a moving source, the frequency heard by a stationary observer is different than the frequency emitted: if the source approaches the receiver, the sound is higher-pitched; if the source moves away from the receiver, the sound is lower-pitched. You hear this same high-pitch, low-pitch pattern as a train goes by, or a race car, or an airplane. The faster the sound source moves, the more noticeable the frequency shift.

The Doppler effect also works for light. If a source of light approaches an observer, the light is shifted toward a higher-frequency end of the spectrum, referred to as a blue shift—a shift toward the blue end of the spectrum; if the source is receding, the light is shifted to a lower frequency, called a red shift—a shift toward the red end of the spectrum. (Think of the visible spectrum as red, orange, yellow, green, blue, and violet, with red as the lowest frequency and violet as the highest.) Since our experience doesn't include extremely fast speeds such as the speed of light, the Doppler effect for light is not noticeable. But, if the amount of frequency shift is measured, the speed of the source can be calculated. Doppler radar is used by weather forecasters to obtain the speed of frontal systems and by baseball observers to find out how fast a pitch travels. Applied to astronomy, the Doppler effect allows the determination of the speed of stars or star groupings. This is what V. M. Slipher did for nebulae that caused such a stir at the American Astronomical Society meeting.

BACK TO HUBBLE

The newly minted American Astronomical Society member and neophyte graduate student Edwin Hubble witnessed V. M. Slipher's presentation on the velocities of faint nebulae. The fact that most nebulae were moving away from us and at extremely high speeds impressed everyone there. The nature of these nebulae was still an open question, as was the size of the Milky Way Galaxy and the size of the whole universe. The cutting-edge position of nebulae on the future of astronomy held tremendous appeal for Hubble.

The next two years at the Yerkes Observatory played right into Hubble's interest. Yerkes was short of staff, having lost several astronomers to the new telescope being built by George E. Hale, the hundred-inch device at Mount Wilson (much more about that later). The other crippling blow was that the director, Edwin B. Frost, suffered from cataracts. He couldn't use the telescope at all and even had to have his mail read to him by Hubble and the other graduate students. Besides taking his turn using the forty-inch refractor in velocity studies, Hubble had almost total use of a twenty-four-inch reflector telescope, and he collected data for his dissertation, "Photographic Investigations of Faint Nebulae." Along the way, he detected a bulge in a nebula NGC 2261. (*NGC* stands for *New General Catalogue of Nebulae and Clusters of Stars*, one of many star catalogs, about which, more later.) This provided nice material for an article in the prestigious *Astrophysical Journal*.

Toward the end of 1916, things began to move very rapidly for Hubble. He had enough data to finish his thesis, and he began writing it. Then, a job offer came from George Hale to work at Mount Wilson, contingent on completing his degree. Next, the United States entered World War I. Hubble hurried his thesis along (perhaps a little too quickly) and volunteered for the war effort. Frost was running out of funding to support Hubble anyway, so he recommended that he take the Mount Wilson job as he helped expedite the thesis. Hubble then wrote to Hale, told him of his service plans, and asked him if the job would still be his after the war. Hale said yes.

In May 1917, Hubble entered the US Army Reserve Officer training program at Fort Sheridan, on Lake Michigan. He chose the infantry and was requested to teach marching by the stars. In August, Hubble was awarded his captain's bars and ordered to active duty. He was assigned to Camp Grant in Illinois and was made commander of the 2nd Battalion, 343rd Infantry Regiment. Training continued during the fall and brutal winter. In January 1918, Hubble was promoted to major, and in July, he was examined and found fit for overseas duty. Deployment to Europe didn't happen until September, and the slow, stormy crossing was extremely difficult, with the threat of U-boats always present. From England, quite different from when Hubble had last seen it almost five years earlier, the 343rd was ferried across the English Channel to France. After training at combat schools, the unit was almost ready to join the war, but the Germans were pushed back and surrendered in October. Hubble said, "I barely got under fire and altogether I am disappointed in the matter of the war."[7] Hubble served several months with occupation forces, then wound up in Cambridge. There, he sat in on a class taught by Arthur Eddington (see chapter 10) and made the acquaintance of astronomer H. F. Newall, who proposed Hubble for membership in the Royal Astronomical Society. In June, Hubble wrote Hale that he was almost ready to return home. Hale responded, "Please come as soon as possible, as we expect to get the 100-inch into commission very soon, and there should be abundant opportunity for work by the time you arrive."[8] Hubble sailed in August, mustered out of the army in San Francisco, and stopped for a short visit to Lick Observatory, near San Jose. The astronomers were a bit thrown by his accent and uniform, and addressed him as "Major Hubble" from then on. He continued traveling south and became an official staff member of the Mount Wilson Solar Observatory on September 3, 1919, probably still wearing his uniform.

In short order, Hubble met the Mount Wilson staff, two of whom would exert tremendous influence on him. One became an invaluable collaborator, the other had personal issues with Hubble (and vice versa) but still assisted his efforts greatly. Let's take the collaborator first.

MILTON HUMASON (1891–1972)

Milt Humason (1891–1972). AIP Emilio Segre Visual Archives, Brittle Books Collection.

Born in Dodge City, Minnesota, Milt (as he was known) didn't progress in school past eighth grade. After a summer camping experience on Mount Wilson, he fell in love with the mountain and convinced his parents to let him stay there for a year. He never returned to school in Minnesota. Milt became a mule driver and hauled materials up Mount Wilson for the observatory. After taking a year off to be a ranch hand, Milt returned to Mount Wilson as a night watchman. Thanks to solar observer Seth Nicholson's instruction, Humason learned how to use the equipment and some mathematics. From there, Humason became a night assistant, helping astronomers. His patience and technical skills with the instruments impressed everyone. In 1919, Hale appointed Humason to the scientific staff as an assistant astronomer. Shortly after his appointment, he met Hubble. Years later, Humason wrote about that first meeting, "'If this is a sample of poor seeing conditions,' Hubble said, 'I shall always be able to get usable photographs with the Mount Wilson instruments.' He was sure of himself—of what he wanted to do, and how to do it."[9] The other night assistants were also appreciative of Hubble's direct approach. In Humason's words, "You knew where you stood with him."[10] Humason learned about Doppler shifts from Slipher and developed his own techniques to make long exposures and measure the velocities of dim nebulae. This took care of one of the variables on Hubble's chart. The other variable, distance, is much more complicated and involves the other Mount Wilson astronomer, with whom there were issues.

HARLOW SHAPLEY (1885–1972)

Harlow Shapley was born on a farm in Nashville, Missouri. After dropping out of school in fifth grade and receiving home schooling, he became a crime reporter for a local newspaper. He then completed a six-year high school equivalency program in two years, he graduated as valedictorian in a class of three, and went to the University of Missouri to study journalism at age twenty-two. The School of Journalism opening was postponed for a year, and Shapley decided to study

Harlow Shapley (1885–1972). From Wikimedia Commons, user Doctree.

the first course he came across in the college catalog. That was archaeology, but Shapley said he couldn't pronounce it, so he settled for the next course, astronomy. After graduation, Shapley was awarded a fellowship to Princeton University, where he earned his PhD studying under Henry Norris Russell. With his brand-new degree, Shapley was hired by Hale and arrived at Mount Wilson in 1914. Shapley used the sixty-inch telescope to study globular clusters (roughly spherical clusters of stars). After Hubble arrived in 1919, both Shapley and Hubble shared the same equipment, but their personalities clashed. Both were Missourians by birth, but that was about all they had in common. Shapley was a pacifist and more of a down-home country boy, with good social skills and a bold manner. The assistants all addressed him as "Doctor." Hubble was almost a polar opposite. His military background was obvious, his British accent and dress were pronounced, and his reports were always conservative in tone. Hubble was a bit standoffish and was called "Major." Privately, Shapley sometimes referred to Hubble as "rubble," and said "he was a Rhodes scholar and he didn't live it down."[11]

As it turned out, they weren't Mount Wilson colleagues for long, but their professional accomplishments were deeply intertwined. Understanding this development and its impact on Hubble's major work will require us to start a bit further back and meet some interesting characters.

HENRY DRAPER (1837–1882) AND
MARY ANNA PALMER DRAPER (1839–1914)

Henry Draper (1837–1882). From Wikimedia Commons, user Jbarta.

Mary Anna Palmer Draper (1839–1914). Courtesy of Special Collections, University Library, University of California Santa Cruz, Lick Observatory Records.

Henry Draper was the son of John William Draper, famous for taking the first photograph of the moon. Henry graduated from New York University (NYU) at age twenty, then indulged his fascination with astronomy by traveling to Ireland to see the largest telescope in use at that time. After becoming a doctor, a professor, and a dean at NYU, Draper took photographs of astronomical objects and their spectra to continue his hobby.

In 1867, Draper married Mary Anna Palmer, a wealthy socialite. She embraced his hobby enthusiastically by working as his lab assistant.

Unfortunately, Draper died at age forty-five of double pleurisy. His wife carried on his work by donating to the Harvard College Observatory to compile a star catalog in her husband's honor. Currently, the Henry Draper Catalogue contains the spectra of more than a quarter million stars, each classified with an HD number. How did the Henry Draper Catalogue get so many stars? It wasn't easy.

THE HARVARD COLLEGE OBSERVATORY

Edward Charles Pickering (1846–1919) was a physicist who was appointed director of the Harvard College Observatory in 1877 to the surprise of many observers. Astronomically, Pickering had no observational experience, but he represented "new astronomy" in that the methods of physics were now being used to investigate stellar structure and evolution. The observatory used "dry plate" photography and an invention of Pickering's own, a meridian photometer, which spread out the spectrum of stars using a calcite prism. Dry plate photography preserves the images on easily viewed and stored glass plates. Harvard also maintained an observatory at Arequipa, Peru, where they observed stars and galaxies such as the Large and Small Magellanic Clouds, visible from the Southern Hemisphere. As the glass photographic plates began to pile up, Pickering started to wonder about his staff's competence to handle this large amount of information. In exasperation, he told one assistant that his housekeeper could do better work. So Pickering fired the assistant and hired his housekeeper, Williamina Fleming. This proved to be a master stroke. Not only was Fleming better, but Pickering only paid her twenty-five cents per hour—comparable to modern minimum wage. The smaller salaries of the women "computers" (they made calculations by hand) made Anna Mary Palmer Draper's money go much further, so Pickering hired women, more than eighty, eventually. They were known as "Pickering's Harem" and proved extremely adept at the required calculations, classifications, and cataloging.

Many of the women on Pickering's staff went on to careers in astronomy and advanced the cause of women in science. That would be an entertaining

Edward Charles Pickering (1846–1919) and his staff, called his "harem."
From Wikimedia Commons, user MarmadukePercy.

story in itself, but the consequences are even more far-reaching. One of Pickering's "computers" in particular made a giant impact on astronomy in general, handing Hubble a tool he desperately needed for his major work. Her name was Henrietta Swan Leavitt (1868–1921).

Leavitt discovered astronomy in her senior year at the Society for Collegiate Instruction of Women, later called Radcliffe College. After remaining in college one more year to study astronomy, Leavitt traveled widely in the United States and Europe, and progressively lost her hearing. Several years later, she volunteered to work at Harvard, and proved to be such a keen observer that she was given the position of chief of the photographic photometry department and was assigned the most difficult task of analyzing variable stars. These were referred to as Cepheid variables, since the first one was found in the constellation Cepheus. She analyzed the

Henrietta Swan Leavitt (1868–1921). From Wikimedia Commons, user Ogrebot.

Magellanic Clouds (neighbor galaxies of the Milky Way visible from the Southern Hemisphere) and found 1,777 new variable stars. By comparing different photographs of the same star, Leavitt was able to establish that the periodicity of variation (from bright to dim and back again) was related directly to the luminosity (brightness) of the star. This period-luminosity relation was expressed mathematically by Ejnar Hertzsprung and extended astronomy's grip on a perennial problem: distance.

ASTRONOMY'S DIRTY LITTLE SECRET: HOW FAR AWAY IS THAT STAR?

Seeing a star, whether with naked eyes or through a telescope, is a good start, but the star's distance from us is another matter. As we look at stars overhead, our usual sense of depth perception fails us. All stars appear to be located the same distance away—far. Because our two eyes look at an object from slightly different positions, each eye sights along its own angle. This phenomenon is called parallax and is used by surveyors to make accurate distance determinations. Because of the small separation between our eyes, they cannot be used to judge long distances very accurately.

Astronomy's simplest technique for determining distances to celestial objects is based on parallax, but it uses a much longer baseline than the distance between our eyes. If the same star is observed at the beginning and end of a six-month interval, it is seen at two different angles (just as our eyes see a distant object from two different perspectives). Measuring the difference in angle and knowing that the baseline of a triangle is the diameter of Earth's orbit enables the distance to the star to be calculated using trigonometry. This was first accomplished by Friedrich Bessel in 1838, when he measured the distance to the star named 61 Cygni.

There are more than three hundred stars within thirty light-years (the distance light travels in one year, almost six trillion miles) of Earth, so we can obtain the distances to these closest neighbors by parallax. As you might expect, other astronomical distances are so large they need an expanded scale. Other galaxies are found at distances of thousands of light-years, millions of light-years, or even farther. However, stars beyond about a hundred light-years are so far away that our telescopes are unable to measure their angles accurately enough to determine their distance.

This is where Leavitt's period-luminosity comes into focus. In 1913, Danish astronomer Ejnar Hertzsprung used this relationship in a very clever way. Measuring the period (time from one bright flash to the next) allowed him to determine the star's intrinsic luminosity, then the measured apparent luminosity allowed the distance to be determined using the inverse square law. (For example, starting with two equally bright stars, the one twice as far away would appear only one-fourth as bright.) This technique was then used by other astronomers to measure more stellar distances, as long as they could spot the particular kind of star needed—a Cepheid variable.

SHAPLEY AGAIN

Harlow Shapley made excellent use of the new measuring tool almost immediately. He studied globular clusters (a spherical collection of stars, similar in shape to a globe) and found enough Cepheid variables within them to determine their distances. (The stars he thought were Cepheid variables were not classic ones and had a slightly different period-luminosity relation, which caused Shapley to overestimate the size of the galaxy.)

This brings up a sore point in astronomical history. In the early 1900s, there were real live questions in astronomy. What is the size of the Milky Way Galaxy; how big is the universe; what are those pesky nebulae, and how far away are they?

We need some history here to put these questions into perspective. The term *galaxy* is derived from the Greek term *galaxias kyklos*, which translates to "milky circle" (Milky Way). Swedish philosopher Emanuel Swedenborg (1688–1772) theorized that all stars formed one large group, with the solar system just one part. In Swedenborg's book *Principia Rerum Naturalium* (1734) he proposed that our solar system of sun and planets was formed from a rapidly rotating nebula. The source of Swedenborg's information wasn't any scientific observation, although he did study science. He got his information from a séance that allegedly included visitors from heaven. Later visions encouraged Swedenborg to reveal theological information, and a religion eventually sprang up from his teachings.

The galaxy story continues with an Englishman, Thomas Wright (1711–1786) of Durham, who built scientific instruments and model solar systems that he sold to the nobility. In his 1750 book *An Original Theory or New Hypothesis of the Universe*, Wright proposed that stars in the Milky Way are distributed in a kind of shell or disk. He declared, "I can never look upon the stars without wondering why the whole world does not become astronomers."[12] As a scientific instrument maker, he undoubtedly had access to telescopes. However, Wright published no astronomical observations. His book also dealt with religious matters such as the physical location of God's throne.

Stranger still, a review of Wright's book in a Hamburg journal caught the eye of the brilliant philosopher Immanuel Kant (1724–1804). Although Kant misread the account of Wright's work, he proceeded to extend it in a constructive direction. In 1755, Kant proposed that the Milky Way was a lens-shaped disk of stars, rotating about its center. Further, he suggested that the fuzzy patches of light referred to as nebulae were actually systems of stars similar to the Milky Way but very far away. He referred to them as "Island Universes."

So, the beginnings of astronomy's analysis of galaxies came from a philosopher, a theologically oriented instrument maker, and another philosopher. There were other ideas later, but let's cut to the chase. In the 1900s, observational astronomers had generated huge amounts of information but needed some theoretical framework to give it order. Harlow Shapley stood ready to take on that task, as his globular cluster distance measurements provided him with information that implied a new picture for the galaxy. He plotted the locations of the globular clusters in three dimensions and presumed this defined the outer boundaries of the Milky Way. This was a radical new idea, far different than the more traditional galaxy size.

THE GREAT DEBATE

The contrast between Shapley's model of the Milky Way and the more conventional one came into sharp focus in 1920 at a meeting of the National Academy of Sciences in Washington, DC. The young Harlow Shapley was invited to give the William Ellery Hale (George's father) Lecture that year. But, rather than a straight presentation, the lecture was set up as a debate. Shapley was joined by the Lick Observatory's Heber D. Curtis (1872–1942), who had just completed a survey of spiral nebulae. The title of their debate was "The Scale of the Universe."

Heber Curtis (1872–1942). Courtesy of Special Collections, University Library, University of California Santa Cruz, Lick Observatory Records.

Curtis argued the standard view of the time: the Milky Way was lens-shaped, about thirty thousand light-years in diameter, and the sun is located near its center. In his concluding remarks, Curtis departed from the confines of the stated topic and ventured that spiral nebulae are quite distant and constitute separate galaxies. Curtis had no evidence to support this conjecture, but he challenged Shapley to give his opinion.

Shapley gave a much less technical talk than Curtis. His audacious estimate of the Milky Way size was three hundred light-years in diameter, with our sun far from the center. Although Shapley was not prepared for the spiral nebulae issue, he believed they were small gas clouds still within the large confines of the Milky Way Galaxy. He cited recent observations of a Mount Wilson colleague (and personal friend), Adriaan van Maanen (1884–1946). Van Maanen had reported that he measured rotational speeds of the Pinwheel Nebula that would lead to star velocities greater than the speed of light if the nebula was located outside the galaxy. Curtis dismissed van Maanen's work as being unsubstantiated. Later, van Maanen's reports turned out to be faulty.

While the debate had no clear winner and wasn't even well-attended, the idea of a larger Milky Way with the earth far from its center seemed to catch the public's attention. Back at Mount Wilson, Shapley's colleague Edwin Hubble made no secret of his sympathies for Curtis, but it was clear that more information was needed.

There was a hidden agenda to the debate, however. The previous year, Harvard College Observatory's director Edward Charles Pickering had died. Pickering had built Harvard to such prominence that his replacement would have a plum job. Shapley wanted it. Not only did Harvard have excellent staff and equipment, it also had access to Southern Hemisphere observations, so Shapley could explore the Large and Small Magellanic Clouds. After some negotiations, Shapley was offered the directorship by Harvard president Abbott Lawrence Lowell (Percival Lowell's brother), and he took it, effective April 1921. This left Hubble at Mount Wilson with more hundred-inch telescope time, and Shapley filling the job he dreamed about. There had been rumors on Mount Wilson about Shapley's leaving, and they were right. A spot of luck, Hubble might say.

TO WORK

With Shapley no longer competing for telescope time, Hubble went to work with renewed vigor, training the hundred-inch on his favorite targets: nebulae. Within a year, Hubble had surveyed all the nebulae he could find and proposed the beginnings of a classification scheme for nebulae. Later additions and refinements to this scheme produced much controversy, but a far more sweeping result was on the horizon. While Hubble continued to pile up data to refine his scheme, he noticed something extremely unexpected. In October 1923, he found a variable star in M31, the Andromeda Nebula. At first, he thought he had sighted a nova, but subsequent observations showed it to be a Cepheid variable. The hundred-inch had presented him with a means of finding the distance to the Andromeda Nebula. Using Leavitt's period-luminosity relation in the same fashion as Shapley had for globular clusters, Hubble found the distance: almost a million light-years. This settled the debate question about nebulae, and not in Shapley's favor. But Hubble, being the careful fellow that he was, needed more data. By February 1924, Hubble had enough to write to Shapley. "You will be interested to hear that I have found a Cepheid variable in the Andromeda Nebula (M31). I have a feeling that more variables will be found by careful examinations of long exposures. Altogether next season should be a merry one and will be met with due form and ceremony."[13] Cecilia Payne, soon to be Harvard Observatory's first PhD in astronomy, was in Shapley's office when the missive arrived. He remarked to her, "Here is the letter that destroyed my universe."[14] Indeed, this established nebulae as galaxies in their own right, but Hubble never stopped calling them nebulae.

    Bonus Material: Hubble/Shapley Internet interview. See To Dig Deeper for details.

LIFE PARTNER

By the 1920s, Hubble had accomplished quite a lot. He held athletic records in Illinois, he had been a Rhodes Scholar, he had served in World

War I and achieved the rank of major, he had earned a PhD in astronomy, and he had a job as an astronomer at the largest telescope in the world. He had passed his thirtieth birthday and was looked at as an extremely eligible bachelor. That was just about to change.

Grace Burke Leib visited Mount Wilson with friends in 1920 and met Hubble purely by accident. She came from a wealthy family. Her father was a vice president of the First National Bank in Los Angeles. Grace majored in English at Stanford University, where she managed to maintain straight As and become Phi Beta Kappa. Six months after graduation, she married Earl Warren Leib, who came from the most prominent family in San Jose. Just a year older than Grace, Earl had graduated from Stanford with a degree in geology and mining. The childless couple lived with her parents, and he worked in the mining industry.

In 1921, Leib traveled to Amador County, southeast of San Francisco, to obtain mine samples for his employer, the Southern Pacific Company. He climbed down the ladder to the mine shaft, and about halfway down became overcome by gas and fell to his death. Hubble and Grace renewed acquaintances a year after her husband's death, carried on a discreet court-ship, and were married in 1924. In recognition of his meager salary and his new wife's background, Hubble supposedly offered to give up astronomy to become an attorney, but Grace would have none of it. A three-month honeymoon took them to New York, Boston, England, France, and Italy on a trip they found delightful. Along the way, the newlywed Hubbles had picked up several ideas for a home, and not long after their return, they found the ideal setting: San Marino's Woodstock Road. Not far from Mount Wilson, the site had spectacular views, and

Grace Burke Leib Hubble (1889–1981) and Edwin Hubble. Image courtesy of the Observatories of the Carnegie Institution for Science Collection at the Huntington Library, San Marino, California.

was situated on a geologic fault line, a fact that Hubble loved to point out to visitors.

Upon returning to Mount Wilson, Hubble continued to work on his nebulae classification scheme and became embroiled in a controversy with Swedish astronomer Knut Lundmark (1889–1958). During the same time, V. M. Slipher was reaching the limitations of his equipment for measuring radial velocities of nebulae—his telescope was too small. As he tried to measure dimmer nebulae, Slipher needed longer and longer exposure times because his twenty-four-inch telescope just didn't capture enough light. Hubble had less of a problem because of Mount Wilson's hundred-inch telescope's light-gathering power. Besides, Hubble found Cepheid variables in nebulae, so the distance-measuring problem was solved using the work of Leavitt, Hertzsprung, and Shapley. Hubble set Milt Humason to work on the velocities. Humason started by duplicating Slipher's work, confirming all the earlier red shifts. But, as they observed nebulae much dimmer than those used by Slipher, they started experiencing extremely long exposure times. This discouraged Humason, but Hale came to the rescue by obtaining a faster spectrograph and an improved camera. Exposure times decreased from several days to hours, and they began to roll. After many long, cold nights of observing, taking photographs, developing them, and interpreting results, Hubble finally had enough data in March 1929 to publish "A Relation between Distance and Radial Velocity among Extra-Galactic Nebulae" in the *Proceedings of the National Academy of Sciences.*

Hubble wrote Shapley that he had held the paper for more than a year, wanting more data. Two years later, he had the additional data, and published "The Velocity-Distance Relation among Extra-Galactic Nebulae," coauthored with Milt Humason, which included measurements of another fifty nebulae. Ever the cautious skeptic, Hubble handled the theoretical implications with "long tongs." "The present contribution concerns a correlation of empirical data of observation. The writers are constrained to describe the 'apparent velocity-displacements' without venturing on the interpretation and its cosmologic significance."[15]

Strictly speaking, this relationship does apply only to the galaxies

Hubble measured. However, when generalized, it implied something remarkable: The Universe as a whole is expanding.

To see how this happens, consider a simple analogy. Suppose there is a race, the Cosmic Marathon. When the race begins, a few runners take off at 4 miles per hour, some at 3 miles per hour, and others at 2 miles per hour.

One hour into the race, the 4-mph group would have covered 4 miles, the 3-mph group 3 miles and the 2-ph group 2 miles, producing a graph just like the one generated by Hubble. Note that from any runner's perspective, it seems that all others, both the ones ahead and the ones behind, are moving away. That is the point of the Hubble graph: the farther away galaxies move faster; that's how they got to be farther away.

The linear relationship between galaxy recession speed and distance is now called Hubble's law in his honor. Although the distances Hubble determined have been corrected somewhat by modern measurements, Hubble's fundamental results remain valid. The universe consists of galaxies of stars and is huge and expanding.

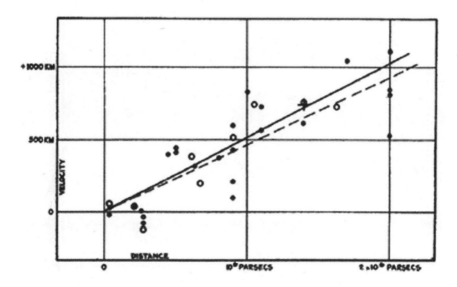

Courtesy of NASA.

EINSTEIN AGAIN

Einstein, Hubble, and others at Mount Wilson, 1931.
Image courtesy of the Observatories of the Carnegie Institution for Science Collection
at the Huntington Library, San Marino, California.

Theoreticians had a field day because there was physical evidence for the ideas of theoreticians Georges LeMaître, Alexander Friedmann, Willem de Sitter, and others. But the happiest theoretician was one who was able to remove a term from his theory: Albert Einstein (see chapters 9 and 10). Now that the necessity for a static universe was removed, the need for Einstein's "cosmological constant" vanished.

In 1931, Einstein came to California. He was scheduled to divide his time between Mount Wilson and Caltech. Sparing no expense, the usually frugal director Walter Adams purchased a "big Pierce Arrow touring car" to convey Einstein up the mountain. Einstein had no experience with equipment of the type on Mount Wilson, and he climbed all over the framework of the hundred-inch telescope, chattering about many details about various instruments—Einstein had done his homework. Elsa Einstein, watching from a safe distance, was told that the giant telescope enabled astronomers to determine the universe's structure. She replied "Well, well, my husband does that on the back of an old envelope."[16]

While he was visiting Mount Wilson, Einstein announced the removal from his equations of the cosmological constant, which he regarded as a blunder. This catapulted Hubble further into the limelight. The *Springfield Missouri Daily News* read, "Youth Who Left Ozark Mountains to Study Stars Causes Einstein to Change His Mind."

The Einsteins soon scheduled a return visit to Pasadena, and arrived there in November 1931. Grace Hubble served as an unofficial hostess and drove Einstein to his commitments. Although he was mostly silent, he once told her, "Your husband's work is beautiful—and he has a beautiful mind."[17] Interestingly, Hubble's attitude about religion was quite similar to Einstein's. When asked about his beliefs by a depressed friend, Hubble said, "The whole thing is so much bigger than I am, and I can't understand it, so I just trust myself to it: and forget about it."[18]

Thanks to the Einstein and Hubble publicity, Mount Wilson turned into a tourist destination, and it became harder to work there. The Hubbles enjoyed the publicity and entertained many Hollywood stars, including Anita Loos, Harpo Marx, Charlie Chaplin, Paulette Goddard, Lillian Gish, Helen Hayes, Frank Capra, Jane Wyatt, George Jessel, Clifford Odets,

Leslie Howard, William Randolph Hearst, and Clare Boothe Luce. Hubble became involved in the design of the two-hundred-inch telescope on Mount Palomar and was granted first use in 1949. His health deteriorated, and he died in 1953.

In 1990, the Hubble Space Telescope was launched, with a primary mirror diameter of ninety-three inches, almost as large as the Mount Wilson telescope of one hundred inches. Because of its position in orbit, there was no atmospheric interference and day/night problems were no longer an issue. The HST offered a phenomenal window into the universe.

Horsehead Nebula. NASA, ESA, and the Hubble Heritage Team (STScI/AURA).

The James Webb Telescope, scheduled for launch in 2018, features a 242-inch mirror and will allow observation of some of the most distant objects in the universe.

Now that we've seen how science can reveal fascinating things about the world of the very large, the next chapter will explore the incredibly small—but with calamitous technological as well as ethical consequences.

CHAPTER 12

DISASTROUS CONSEQUENCES OF LISE MEITNER AND OTTO HAHN'S DISCOVERY OF NUCLEAR FISSION

Hiroshima mushroom cloud. Courtesy of Isabella lynn Lee, from Wikimedia Commons.

WEAPONS OF MASS DESTRUCTION

What did you expect as a follow-up to Hubble? Not weapons. And it was certainly not what the scientists expected, either. But it brings up a critical issue.

Scientists who hypothesize and experiment to deepen our understanding of the universe aren't the only people on this planet—not by a long shot. When utilized by other people science's ideas exert an influence on the human condition. Sometimes the application of science is relatively minor—such as the use of non-polar Teflon molecules to keep polar grease molecules from sticking to pans. Sometimes it develops slowly, like the two hundred years from Bernoulli's principle to the Wright Brothers' airplane. But in this case, the application of science to technology was both huge and quick. While it started from the extremely small, it grew to be enormously significant in only seven years, so hang on.

WHAT DID YOU DO IN THE WAR?

We know what Edwin Hubble did during World War I. Mostly, he trained in the United States, then spent a little time in France, where he earned his major's oak leaf. Did he think about the big picture of astronomy? Possibly.

For contrast, consider James Chadwick (1891–1974). Just before the war, Chadwick studied radioactivity under Ernest Rutherford at the University of Manchester in England, earned his MSc, then went to Berlin to study under Hans Geiger (of Geiger counter fame). When war broke out, Rutherford was sent to an internment camp in Ruhleben, where he remained for the duration of the war. Sounds pretty grim, eh? Not at all. With the cooperation of prison guards, Chadwick was allowed to set up a laboratory in a stable. He was able to obtain quantities of commercially available toothpaste that contained radioactive material, and to set up experiments with toothpaste and gum wrappers that contained aluminum foil.

When the war ended, Hubble worked on the universe's biggest challenges, using Mount Wilson's hundred-inch telescope; Chadwick returned to England to work on the smallest puzzles, namely, the nucleus, by following Rutherford to the Cavendish Laboratories at the University of Cambridge. Both Hubble and Chadwick were quite successful. Hubble's discovery of the expansion of the universe in 1929 revised the thinking of astronomers and opened whole new avenues of large-scale research. In 1932, Chadwick was able to demonstrate the existence of the subatomic particle called the neutron, the second of the components of atomic nuclei, and provided an extremely valuable new tool for probing the atom.

THE NEUTRON: SUSPECTED BUT VERY ELUSIVE

Rutherford had found evidence for the existence of the nucleus of the atom when he set up an experiment in which thin sheets of gold foil were bombarded by alpha particles, naturally occurring particles in radioactive decay of certain elements (see the introduction). In 1920, he showed that the simplest atom of all, hydrogen, had only one positively charged particle in its nucleus. He named this particle the proton. On the other hand, heavier nuclei in the periodic table had masses that weren't simply multiples of their number of protons. Further, radioactive decay of some elements gave off particles other than the alpha. One of these turned out to be the electron.

What was a tiny electron doing in the nucleus with all its massive protons? Was there some massive particle in the atom's nucleus in addition to the proton, but with no charge to upset the atom's equal numbers of electrons and protons? We need some background here before we can understand Chadwick's work.

RUSH TO DISCOVERY

Almost as if it was a deadline-beating phenomenon, the years before the turn of the twentieth century featured fascinating last-minute discoveries:

1895: X-rays, discovered by Wilhelm Röntgen
1896: radioactivity, discovered by Antoine Becquerel
1897: the electron, discovered by J. J. Thomson

While each of these discoveries was interesting in its own right, let's just look at radioactivity to understand how it all shakes out:

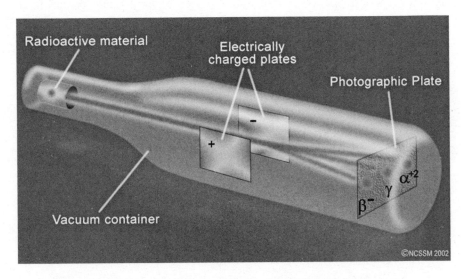

Radioactive decay products. Created by North Carolina School of Science and Mathematics. Licensed CC BY-NC-SA 4.0, used with permission.

- alpha (α)-particles are actually two protons and two neutrons stuffed together (charge = +2) and can be stopped by a shield as flimsy as a piece of paper; that's why Rutherford's experiments used extremely thin gold foil, since it could be made wafer-thin.
- beta (β)-particles are actually electrons (charge = -1) and cannot penetrate thin metal.
- gamma (γ)-particles are actually high-frequency light radiation (charge = 0) and can penetrate a thick lead shield.

While Hubble had used the world's biggest telescope to probe the large world of the nebulae by sensing visible light the nebulae emitted, experimenters investigating the nucleus had a much different problem. They had to shoot tiny invisible particles from radioactive decay and use cleverly designed detectors to find out what happened after these probes collided with the nucleus. Their results were confusing but interesting. Beta and gamma particles revealed a bit of information, and alpha particles were massive enough to penetrate into the heart of the atom, but their positive charge kept them from entering the nuclei because they were repelled by the positively charged nucleus. A new probe was needed.

BACK TO JAMES CHADWICK

Suspicion and reality came together in 1932. James Chadwick, in addition to serving as Rutherford's assistant director of research, managed to isolate and identify the particle that completed their understanding of the nucleus. This particle was slightly more massive than the proton and carried no electrical charge. Rutherford had already hypothesized the particle and named it the neutron. For his efforts, Chadwick was awarded the Nobel Prize in 1935. Just in case anyone might think the cartoon character Jimmy Neutron was named after Chadwick, it is true that Jimmy first appeared merely twenty-five years after Chadwick's death.

Besides completing the ideas about the atom's constituents, the neutron was almost immediately recognized as a valuable research tool. It had sufficient mass to penetrate an atom, and its electrical neutrality allowed it to penetrate the nucleus without being repelled. In short order, four different laboratories used neutrons to bombard heavier elements in order to produce elements with more neutrons in their nuclei. Technically these new versions of the original element are referred to as isotopes of that element because they have the same chemical properties as other isotopes of the same element, so they all occupy the same spot on the periodic table.

COMPETITION

Let's look at the competition between various laboratories as scientists endeavored to complete the understanding of the atomic nucleus.

The Cavendish Laboratories in Cambridge, England: The discoverer of the neutron was on staff here, and so the lab had a head start on the others.

The Radium Institute in Paris: Maria Skłodowska Curie's daughter Iréne Joliot-Curie and her husband, Frederic Joliot, produced radioactive isotopes by bombarding boron, magnesium, and aluminum with α-particles. For their work, they were awarded the Nobel Prize in Chemistry in 1935. When the couple tried neutron bombardment of uranium, the results were curious and confusing.

The University of Rome: Enrico Fermi's group had the most efficient neutron source and set out on the most ambitious program. Their goal was to bombard every element with neutrons and check for resulting radioactive isotopes. One of Fermi's innovations was to slow the neutrons down to facilitate their capture by nuclei. All went smoothly until they reached the largest naturally occurring element, uranium, atomic number 92. Colleagues thought he might have created a new element, one with atomic number 93, but Fermi was not so sure. In Fermi's words, "We did not know enough chemistry to separate the products from one another."[1]

In 1934, German chemist Ida Tacke Noddack criticized Fermi's results because Fermi's group hadn't tested their products for elements lighter than lead. "It is conceivable that the nucleus breaks up into several large fragments, which would of course be isotopes of known elements but would not be neighbors of the irradiated element."[2] Since these projects were not in her direct field, she had no way to pursue them, but later developments were exactly along the lines she suggested.

Fermi won the 1938 Nobel Prize for his efforts. As you may recall from the chapter 10 discussion of Maja Einstein Winteler's decision to leave Italy because of anti-Semitic laws, Italy was becoming an unsafe place for Fermi's Jewish wife, Laura. The Fermis convinced the Mussolini government to let the whole family go to Sweden to collect the Nobel

Prize. Once they got the award, they didn't return to Italy. Instead, they traveled directly from Stockholm to New York, arriving January 2, 1939. As they landed, Enrico said to Laura, "We have founded the American branch of the Fermi family."[3]

The Kaiser Wilhelm Institute for Chemistry in Berlin-Dahlem: The team of Lise Meitner and Otto Hahn had been collaborating since 1907 and wound up with the most complete understanding of uranium bombardment of any of the groups. The connections get complex, so let's have a look at each member of the team before we see their enormously significant results.

OTTO HAHN (1879–1968)

Otto Hahn was the youngest of four children. As a youth, Hahn loved chemistry, he even performed experiments in the family laundry room. Although his father wanted him to study architecture, his desires for chemistry won out, and he wound up with a PhD in chemistry in 1901 from the University of Marburg. After postdoctoral studies at University College in London and at McGill University in Montreal with Ernest Rutherford, Hahn had found a specialty: radiochemistry. He managed to isolate several radioactive isotopes, and he received this comment from Rutherford: "Hahn has a special nose for discovering new elements."[4]

Later in 1906, Hahn returned to Germany to work with Emil Fischer at the University of Berlin, and, within a few months, he discovered two new radioactive isotopes using extremely unsophisticated equipment. A year later, Hahn met the physicist Lise Meitner. They began a lifelong friendship and an extremely fruitful scientific collaboration.

THE PARTNERSHIP'S OTHER HALF:
LISE MEITNER (1878–1968)

As the third of eight children of wealthy parents, Lise (originally Elise) Meitner was tutored to bring her education level up to that of her brothers.

She excelled in physics and was admitted to the University of Vienna, where she was inspired by Ludwig Boltzmann. Her nephew, Otto Robert Frisch, wrote that "Boltzmann gave her the vision of physics as a battle for ultimate truth, a vision she never lost."[5]

In 1905, Meitner became the second woman to earn a PhD from the University of Vienna. Her thesis discussed both theoretical analysis and experimental determinations, which were a hallmark of her entire career. Thanks to family financial backing, Meitner went to Berlin and was allowed to attend Max Planck's lectures, a most uncommon occurrence for a woman. A year later, she became Planck's (unpaid) assistant and sought a place to perform laboratory experiments. Meitner approached Professor Heinrich Rubens, who offered her space in his physics lab but mentioned that chemistry professor Otto Hahn had indicated an interest in collaborating with her. The shy and reserved Meitner thus had a choice between working with a substantially older physicist or a chemist of her own age. She had worked with radioactivity at Vienna, so she had a headstart in that area. She chose to work with Hahn.

An early difficulty was that women weren't allowed in the chemistry labs, allegedly because of safety—"their hair might catch fire." One

Lise Meitner and Otto Hahn. From Wikimedia Commons, user Pieter Kuiper.

wonders about some of the men with their full beards. The solution was to use a basement room, which had been a former carpenter's workshop with an outside entrance. A restaurant down the street provided the nearest toilet, but Meitner was determined and stuck it out. With Hahn performing the tedious chemical separations and Meitner making the physical measurements and analyzing the physics, the team discovered several new radioactive isotopes and published two papers on β-particles.

Within the next few years, many changes occurred, including the onset of World War I, the founding of a new institute, and the awarding of a full professorship for Meitner at the University of Berlin. In addition, Meitner connected with many other physicists of the day. For example, she taught a seminar with Leo Szilárd in 1930 and corresponded with Albert Einstein about physics problems. Einstein was impressed. He referred to Meitner as "Our Marie Curie."[6]

FINALLY, NEUTRON BOMBARDMENT

Although Meitner and Hahn's neutron source wasn't as efficient as Fermi's, they had Hahn's excellent skill with the chemistry as a bonus. Once an element is bombarded with a neutron, the product must be analyzed to ascertain which isotopes are actually present. Soon, the team was augmented by the addition of Fritz Strassmann, who had resigned from the Society of German Chemists when it was taken over by Nazis in 1933. Although Strassmann was blacklisted, Hahn and Meitner found a half-pay position for him at the University of Berlin in 1934, and he assisted greatly in their neutron bombardment project. (Strassmann and his wife also concealed a Jewish friend in their apartment for months during the war, at great personal risk.)

The Meitner/Hahn/Strassmann group identified sixteen radioactive isotopes with various half-lives, but they encountered a personal half-life problem close to home. Lise Meitner had been born Jewish but had converted to Christianity in 1908, as had two of her sisters. Further, as an Austrian citizen, she paid little attention to the accession of Hitler as Germany's chancellor. After 1933, many of Meitner's physics colleagues had already fled Germany, including Einstein, Szilárd and even her nephew Otto Frisch. But Meitner ignored political developments and focused on her work.

In fall 1938, several developments occurred in quick succession. The team was reaching the end of their bombardment project by working on uranium, and the Nazis annexed Austria (*Anschluss*). Meitner's friends urged her to escape and arranged to smuggle her to the Netherlands. Otto Hahn even gave her a diamond ring he had inherited from his mother in case she needed to bribe a border guard. She narrowly escaped and eventually wound up in Manne Siegbahn's laboratory in Stockholm, Sweden. (The ring's eventual whereabouts are unknown.) In November, Hahn traveled to Copenhagen. He and Meitner planned another set of experiments involving uranium in their neutron bombardment series. Hahn and Strassmann then carried out the experiments, they but were perplexed by the apparent presence of the much lighter element barium in the product.

In December, Hahn wrote to Meitner, "Perhaps you can come up with some sort of fantastic explanation, we know ourselves that it can't actually burst apart into barium."[7] Meitner advised them to continue with further tests, but their prior results were confirmed. Meitner's physicist nephew, Otto Robert Frisch, who worked with Niels Bohr in Copenhagen, was visiting his aunt over the Christmas holiday, and Meitner showed him the Hahn letter as they hiked together in the snow. With Meitner walking and Frisch on skis, they discussed the result and drew diagrams on the little pieces of paper that somehow seem to collect in scientists' pockets.

Using Bohr's model of the nucleus as a liquid drop, they tried to visualize a neutron causing the wobbly drop to split. Frisch referred to this as fission, since the products were less massive than the original nucleus. Using Einstein's famous $E = mc^2$, they realized that a substantial amount of energy would be released in the process. Meitner wrote back to Hahn, and they agreed to publish separate papers, with Hahn and Strassmann discussing the chemistry, and Meitner and Frisch the physics.

Here's an excerpt from Meitner and Frisch's paper: "O. Hahn and F. Strassmann have discovered a new type of nuclear reaction, the splitting into two smaller nuclei of the nuclei of uranium and thorium under neutron bombardment. . . . It can be shown by simple considerations that this type of nuclear reaction may be described in an essentially classical way like the fission of a liquid drop, and that the fission products must fly apart with kinetic energies of the order of hundred million electron-volts each."[8]

THE WORD SPREADS

Otto Frisch wasted no time in explaining this result to Niels Bohr in Copenhagen, who understood it almost immediately. Not only was a substantial amount of energy released, but more neutrons might be generated, which might lead to more fission events, and more fission events . . . a chain reaction. Bohr arrived in the United States in January of 1939 for a conference at George Washington University on low-temperature physics and superconductivity. Addressing this group, Bohr ignored the stated conference topic and excitedly announced the fission experiment results. Comments were remarkably subdued. One physicist recalls a colleague whispering in his ear, "Perhaps we should not discuss this."[9]

WAY MORE THAN DISCUSSION

A corner had been turned. It was clear that fission produced energy, but the questions of how the energy might be used in a practical sense and when it might be available were unknown. Even more importantly, who might control the energy and for what purpose were critical questions. The initial experiments were carried out in Germany, perhaps putting Germany in the driver's seat. Germany's military ambitions were on clear display, as its troops had just overrun the Sudetenland, and more German expansion was clearly in the wind. Bohr's native Denmark felt the pressure of the Nazi threat: many of those scientists at the conference had already fled Europe to avoid Hitler.

The next six years produced a whirlwind of scientific activity and even more technological and political commotion. To track these developments from a human perspective, we will focus on two individuals who played key roles: Enrico Fermi and Leo Szilárd.

ENRICO FERMI (1901–1954)

As we saw earlier, Fermi was in on the ground floor of neutron bombardment. He won the Nobel Prize in 1938 and promptly sailed for the United States. Fermi and his family arrived in early January 1939, just as the Meitner/Hahn papers were being published. When Bohr arrived in America for the conference, at which he revealed the fission news, he immediately sought out Fermi to tell him first. Fermi saw the possibilities but suggested a go-slow policy until their understanding was more complete.

Enrico Fermi (1901–1954).
From Wikipedia, user Pieter Kuiper.

LEO SZILÁRD

A friend of both Einstein and Meitner from their Berlin days, Szilárd had been involved in many different aspects of physics. He conceived of the linear accelerator, cyclotron, and electron microscope and even filed a few patent applications, but he didn't follow through on any of these. In 1933, Szilárd saw the political handwriting on the wall and left Germany, settling in England. Learning of Ruther-

Leo Szilárd (1898–1964). From Wikimedia Commons, user Panoptik~commonswiki.

ford's bombardment of lithium nuclei, Szilárd envisioned neutrons being given off in the process. These in turn could bombard other nuclei, leading to a chain reaction. He patented this idea and assigned it to the British Admiralty to ensure its secrecy.

Szilárd then worked on radioactive isotopes for medical treatment. Seeing a European war as inevitable, he immigrated to the United States in 1938 and conducted research at several universities across the country, winding up in New York in January 1939. When Szilárd heard about the uranium fission, it brought up the fear of Germany making a bomb using the chain reaction principle. Szilárd had no university affiliation, so he needed some strong academic connection. Fermi had a professorship at Columbia, so Szilárd decided to approach him.

THE ODD COUPLE

It's hard to imagine two people less similar. Fermi was a careful experimenter who always wanted to carry a project through to completion, whereas Szilárd nimbly jumped from one idea to another. Bernard T. Feld, Szilárd's research assistant, describes the difference between Fermi and Szilárd this way: "Fermi would not go from point A to point B unless he knew all he could about A and had reasonable assurances about B. Szilard would jump from point A to point D, then wonder why you were wasting your time with B and C."[10] Fermi was an early riser, sometimes getting up by five am to plan his day. Szilárd often slept in, then soaked in his bathtub for inspiration. Nevertheless, Szilárd was made a guest researcher at Columbia.

He and Fermi began the research that led to a practical nuclear fission reactor. They learned several things fairly quickly: the rare uranium isotope of atomic mass 235 undergoes fission when hit by slow neutrons, as opposed to the far more naturally occurring isotope 238, which cannot support a chain reaction; neutrons can be slowed by several substances, such as "heavy" water (which contains a large proportion of the more massive hydrogen ion, deuterium), or else highly purified carbon in the form of graphite. Both of these developments implied that large quantities of rare and costly materials would be necessary in order to make a fission reactor. Fermi continued his careful experimental work.

Szilárd however took a different route. He approached financiers and tried to raise money for private research. When that approach proved fruitless, he tried an entirely different tack. Starting in July, Szilárd paid several visits to his old friend Albert Einstein, who was vacationing in Peconic, Long Island. On the first visit, Szilárd and a college friend, physicist Eugene Wigner, got hopelessly lost. After going in circles for hours, they decided to ask a sunburnt boy of about seven if he knew where Einstein lived. "Of course I do," he said and pointed the way. When they arrived, Einstein greeted them cordially, dressed in a white undershirt and rolled-up white trousers. When they outlined the fission developments, Einstein answered, "I haven't thought of that at all."[11] But he shared their concern, and the three of them drafted a letter that they planned to send to the Belgian ambassador and the US State Department. After a few more letter drafts and trips to Peconic, the following letter was generated:

From Wikimedia Commons, user Schutz.

The letter was delivered to President Franklin Delano Roosevelt by Alexander Sachs, a personal friend of the president's and a vice president of Wall Street's Lehman Corporation. Roosevelt said to him, "Alex, what you are after is to see the Nazis don't blow us up." "Precisely," said Sachs.[12] The first step in a lengthy sequence was put in place: an advisory committee was set up.

THE MANHATTAN PROJECT

In June 1941, Roosevelt created the Office of Scientific Research and Development (OSRD) headed by Vannevar Bush (no relation to Presidents Bush 41 and Bush 43). In October, Roosevelt approved the fission effort, called the Manhattan Project, and agreed to coordinate with American efforts with those of the British. The British effort, code-named Tube Alloys, had a recent breakthrough investigating the amount of uranium 235 needed to sustain fission. British researchers Frisch and Peierls (more on Peierls later) found that the critical mass (amount required to have a self-sustaining chain reaction) of uranium required was approximately 1 kg (2.2 lbs). Does *Frisch* sound familiar? Otto Frisch was Lise Meitner's nephew, the one who had first named fission.

In December 1941, World War II began in earnest for the United States. Japan declared war on America and attacked Pearl Harbor; Germany declared war on the United States shortly thereafter. The Manhattan Project assumed even greater importance, since most everyone now agreed with Leo Szilárd. It was essential that the Americans beat the Germans to developing the fission bomb.

To maintain the security Szilárd had insisted on from the very beginning, work on the project was split up and carried out in different locations: Hanford, Washington; Oak Ridge, Tennessee; Long Island, New York; Chicago, Illinois; and Los Alamos, New Mexico. The idea was to keep the scientists isolated from each other to minimize communication so that no one could get a view of the entire project. Secrecy was so foreign to scientists' usual procedure that it failed in many instances. Some instances were harmless, such as the efforts of Richard P. Feynman, who routinely opened the combination locks of absent colleagues' safes to obtain needed reports. Feynman then left them a note telling them what he had done. Other security breaches were much more serious, such as the information supplied to the Russians by British physicist Klaus Fuchs.

Eventually, the Manhattan Project employed 130,000 people at thirty different sites in the United States, Canada, and the United Kingdom. Physicist J. Robert Oppenheimer was the director of the Los Alamos

National Laboratory where the bombs were designed and built, and Brigadier General Leslie R. Groves from the Army Corps of Engineers was the overall project director. Fermi had a hand in every scientific aspect of the project. Szilárd was the project gadfly, making comments on every decision, sharing valuable insights, cajoling suppliers, and annoying almost everyone. His major clash was with General Groves. Both men had stocky builds, but the similarity ended there. Szilárd disliked authority and engineers, and Groves was both. Szilárd spoke with a Hungarian accent, had a difficult-to-trace background, and moved about erratically. Oddly, the security issue that so irritated Groves had been a hallmark of Szilárd's efforts. Nevertheless, Groves tried to have him removed from the project as a security risk, but failed when Szilárd turned over more details of his previous life and listed several references, including Einstein.

The first nuclear reactor was constructed in a former squash court under the stands at Stagg Field at the University of Chicago. Called CP-1 (for Chicago Pile-1), the reactor consisted of uranium pellets (six tons plus thirty-four tons of uranium oxide) and graphite blocks (four hundred tons), with control rods of cadmium, indium, and silver. When a uranium nucleus undergoes fission, the neutrons emitted are slowed by the graphite so that they can cause more fissions. Conversely, the control rods absorb neutrons so that they are unable to create more fissions.

On December 2, 1942, the construction of the CP-1 was complete, and all control rods were removed, save one. As the last control rod was withdrawn in increments, Fermi monitored the neutron emissions carefully until, finally, enough neutrons caused fissions to make the process self-sustaining—the pile had "gone critical." After twenty-eight minutes of chain reaction, Fermi ordered the control rods replaced.

The test was a success. Director Arthur Compton sent a cable to James B. Conant, the chairman of the National Defense Research Committee, announcing the result in an ad hoc coded form:

Compton: The Italian navigator has landed in the New World.
Conant: How were the natives?
Compton: Very friendly."[13]

Szilárd's view was a little different. He shook Fermi's hand and said that this would be regarded as a "black day in the history of mankind."[14] Now that a fission reactor had been demonstrated, the probability of building a bomb had become much more likely.

GERMAN NUCLEAR EFFORTS

Kurt Diebner headed the German nuclear fission research projects that began in April 1939. One of the major contributors was Werner Heisenberg, who had earned his PhD under Niels Bohr in Copenhagen. The research proceeded along similar lines as the Manhattan Project, but the Germans had a substantial headstart. By the middle of 1941, the Germans could see a path to building a fission reactor, but they realized it would require an enormous effort, both in personnel and resources.

In September 1941, a curious thing happened. Heisenberg accepted an invitation to speak at a German cultural institute in Copenhagen, in German-occupied Denmark. While he was there, Heisenberg met with Bohr, and they took a long walk together. Since there were no witnesses, many have speculated about the topic of their conversation and its possible relevance to nuclear fission research. Such speculations included a popular book (*Brighter Than a Thousand Suns* by Robert Jungk) and a Tony award–winning play, *Copenhagen* by Michael Frayn.

Whatever was said between the colleagues, Germany's leaders soon made a decision to concentrate the war effort on rockets and jet airplanes. The nuclear research was split up among nine institutes, the directors of which set their own objectives, so no coordinated effort was undertaken.

After the war ended, British intelligence rounded up many German scientists and interned them at Farm Hall, near Cambridge. Their quarters were bugged, and transcripts of the listening device recordings were scanned for information about how far along the German program had gotten. Here's a quote from Heisenberg: "The point is that the whole structure of the relationship between the scientist and the state in Germany was such that although we were not 100 percent anxious to do it, on the other hand we were so little trusted by the state that even if we had wanted to do it, it would not have been easy to get it through."[15]

And here's a comment from Otto Hahn, chiding his colleagues who had worked on the project: "If the Americans have a uranium bomb then you're all second-raters."[16]

AMERICAN FISSION BOMB

Indeed, the Americans did have a uranium bomb in 1945, but its development required more than two years of concentrated engineering and technological development beyond the reactor CP-1 stage. Besides the technical effort, a giant principle came into play here.

Let's look at a much milder example first. In the 1960s, one of us (AW) worked as an engineer for a company in the aerospace industry. The projects were long-range and research-oriented, but as the war in Vietnam heated up, the company investigated a new project outside our usual range. I was still busy on other matters, but a friend of mine was assigned to work on a proposal for a computer system for a new missile system. After learning the details of the project (he remembers a discussion of kill radius along the null range vector), my friend suggested that, rather than use a miniature computer to guide a proposed multiple warhead missile, the military should hire small people. He knew his sarcastic comment would get him either reassigned or fired, but he said it was a matter of ethics, which was more important than the job. My friend got a new assignment, and the project was dropped. No harm, no foul?

On the much larger scale of the devastating effects of the fission bomb, what were the ethical concerns of the scientists, engineers, technicians, military personnel, and their families? Could they, like my aerospace engineer friend, anticipate the harmful consequences of the Manhattan Project? Certainly, opinions and ethics varied all over the map. Further, the multiple sites and secrecy involved prevented any one person at any one time from having a clear picture at any one site of the whole project. But we do know about one person who had a bird's eye view, the ability to jump from A to D in a single bound, and the courage to speak up, to anybody: Leo Szilárd.

LEO SZILÁRD AGAIN

By mid-1943, the Nazis began suffering military reverses on every front. To forward thinkers like Leo Szilárd, the end of the war was in sight, even a year before D-Day. Szilárd was far more interested in nuclear power for peaceful purposes, and he worried about the destructive aspects of the bomb. When he later recalled the events of March 1944, he wrote: "Initially we were strongly motivated to produce the bomb because we feared the Germans would get ahead of us and the only way to prevent them from dropping bombs on us was to have bombs in readiness ourselves. But now, with the war won, it was not clear what we were working for."[17] As usual, Szilárd was thinking far in advance of anyone else, but other people began to catch up.

By early 1945, a rapid succession of events occurred that changed the world scene forever.

- March: Szilárd drafted a letter for Einstein to sign and send to President Roosevelt. The letter introduced Szilárd and requested a meeting between Roosevelt and Szilárd. The letter stated, "The terms of secrecy under which Dr. Szilard is working at present do not permit him to give me information about his work; however, I understand that he now is greatly concerned about the lack of adequate contact between scientists who are doing this work and members of your Cabinet who are responsible for formulating policy."[18]

Used with permission from Sidney Harris.

- April: President Roosevelt died unexpectedly without seeing Szilárd. He was succeeded by Vice President Harry S. Truman, who had no knowledge of the Manhattan Project until after he was sworn in.
- May: Germany surrendered. This was known as VE Day, for victory in Europe. Some workers left the Manhattan Project, but the vast majority stayed to finish. Late in the month, Szilárd had an appointment at the White House. The president's appointments secretary, Matthew J. Connelly, said the president wanted former Senator James Byrnes of South Carolina to handle this matter. Szilárd agreed and traveled to Spartanburg in the former senator's home state, where he delivered his memorandum regarding the moral implications of the nuclear bomb. The meeting was unproductive. Byrnes's recollection was that Szilárd's "general demeanor and his desire to participate in policy-making made an unfavorable impression on me." Szilárd said, "I was rarely as depressed as when we left Byrnes' house."[19]
- June: Arthur Compton, the director of the Met Lab, as the Chicago Manhattan Project site was called, wrote very insightfully about the political activities of his scientific colleagues: "The scientists will be held responsible, both by the public and their own consciences, for having faced the world with the existence of the new powers. The fact that the control has been taken out of their hands makes it necessary for them to plead the need for careful consideration and wise action to someone with authority to act. There is no other way in which they can meet their responsibility to society."[20]
- July 17: Szilárd drafted A PETITION TO THE PRESIDENT OF THE UNITED STATES, which stated, in part:

> In view of the foregoing, we, the undersigned, respectfully petition: first, that you exercise your power as Commander-in-Chief, to rule that the United States shall not resort to the use of atomic bombs in this war unless the terms which will be imposed upon Japan have been made public in detail and Japan knowing these terms has refused to surrender; second, that in such an event the question whether or not to use atomic bombs be decided by you in

light of the considerations presented in this petition as well as all the other moral responsibilities which are involved.

Signed by Leo Szilárd and sixty-nine cosigners.

The petition was routed through official channels and had no discernable effect.

- July 16: Unknown to Szilárd, the first nuclear bomb was tested (code name: Trinity) at Los Alamos, New Mexico, at 5:45 am. Called "the Gadget," the bomb detonated successfully. Military planners continued planning a strike on mainland Japan. President Truman's diary reflects his instructions: "I have told the Sec. of War, Mr. Stimson, to use it so that military objectives and soldiers and sailors are the targets and not women and children."[21]
- August 6: The B-29 *Enola Gay* dropped a single nuclear fission bomb on the Japanese mainland city of Hiroshima at 8:15 am local time. The bomb detonated at an altitude of 19,000 feet and packed the destructive power of 16,000 tons of TNT. Achieving that effect with conventional bombs would require at least 2,000 fully loaded B-29s dropping 500-lb. bombs—simultaneously. The results were horrific: 70,000–80, 000 were killed instantly (20,000 soldiers) and another 70,000 injured. Three days later, a second bomb was dropped on Nagasaki, causing 22,000 to 75,000 deaths.

THE INHUMAN SIDE OF SCIENCE?

Even though the destruction experienced by the Japanese cities closely matched their predictions, many of the Manhattan Project scientists were horrified. The *Bulletin of the Atomic Scientists* was founded in 1945 by Manhattan Project scientists who "could not remain aloof to the consequences of their work."[22] The organization's early years chronicled the dawn of the nuclear age and the birth of the scientists' movement, as told by the men and women who built the atomic bomb and then lobbied with both technical and humanist arguments for its abolition (see more at:

Head-on collision of Physics, Technology, and Ethics.
Used with permission from Sidney Harris.

http://thebulletin.org). On the cover of every issue is a Doomsday Clock, representing the scientists' estimate of the countdown to possible global catastrophe.

IN THEIR OWN WORDS

Here are quotes from some of the people involved:

> ". . . the greatest thing in history."
>
> —Harry S. Truman

Doomsday Clock. Courtesy of Sancheevi Sivakuma, from Wikimedia Commons.

"Japan was at the moment seeking some way to surrender with minimum loss of 'face.' It wasn't necessary to hit them with that awful thing."

—General Dwight D. Eisenhower

"It is my opinion that the use of this barbarous weapon at Hiroshima and Nagasaki was of no material assistance in our war against Japan. The Japanese were already defeated and ready to surrender. My own feeling was that in being the first to use it, we had adopted an ethical standard common to the barbarians of the Dark Ages. I was taught not to make war in that fashion, and wars cannot be won by destroying women and children."

—Admiral William D. Leahy,
Former Chair of the Joint Chiefs of Staff

"I have no remorse about the making of the bomb and Trinity [the first test of an A-bomb]. That was done right. As for how we

> used it, I understand why it happened and appreciate with what
> nobility those men with whom I'd worked made their decision.
> But I do not have the feeling that it was done right. The ulti-
> matum to Japan [the Potsdam Proclamation demanding Japan's
> surrender] was full of pious platitudes. . . . Our government
> should have acted with more foresight and clarity in telling the
> world and Japan what the bomb meant."
>
> —J. Robert Oppenheimer[23]

Since the nuclear bombs were dropped while many German scientists were still interned at Farm Hall, the transcripts reveal some of their reactions. Otto Hahn contemplated suicide, believing himself personally responsible for the many Japanese victims, while less than two weeks after the announcement Werner Heisenberg had figured out the process by which the bomb was built.

"So, you're the little lady who got us into all of this!" said President Harry S. Truman in 1946 when he met Lise Meitner.[24]

When asked to join the Manhattan Project in 1943, Lise Meitner replied, "I will have nothing to do with a bomb!"[25]

In 1947 Einstein told *Newsweek* magazine "Had I known that the Germans would not succeed in developing an atomic bomb, I would have done nothing."[26]

Leo Szilárd sums it up: "I have been asked whether I would agree that the tragedy of the scientist is that he is able to bring about great advances in our knowledge, which mankind may then proceed to use for purposes of destruction. My answer is that this is not the tragedy of the scientist; it is the tragedy of mankind."[27]

In 1947, Leo Szilárd switched from physics to biology and helped found the Salk Institute in La Jolla, California. Szilárd continued his eclectic brilliance in the new field and made many contributions to biological research. His insightful suggestions inspired one of his colleagues to remark that he loved being around Szilárd because "Leo got excited about something before it was true."[28] That colleague was James Watson, who figures prominently in the next chapter.

CHAPTER 13

MAURICE WILKINS, ROSALIND FRANKLIN, JAMES WATSON, AND FRANCIS CRICK DETERMINE THE STRUCTURE OF DNA

God touches DNA. Used with permission from Sidney Harris.

> The greatest single achievement of nature to date was surely the invention of the molecule DNA.
>
> —Lewis Thomas[1]

JAMES WATSON'S DNA NOBEL PRIZE SELLS FOR $4.8M

Nobel medal. Courtesy of Anubis3, from Wikimedia Commons.

Yes, this headline from the December 5, 2014, BBC News involves the same James Watson who loved being around Leo Szilárd because he "got excited about something before it was true." How in the world did Watson get a Nobel Prize for work he did at the tender age of twenty-five? And why did he sell the medal, the first living Nobel winner to do so? The answers to these questions involve several other people, and feature more than a few foibles, misunderstandings, competition among groups, faulty communications, and misjudgments—in other words, the usual suspects.

LEADER OF THE PACK

Let's start with the oldest of the participants in this drama, someone who turns out to be a key player—but at a distance: John Turton Randall (1905–1984)

John T. Randall (1905–1984). From BioMed Central Blogs, *On Biology*, Sept. 2, 2013.

John T. Randall was born in quite modest surroundings in Lancashire, England. At the University of Manchester, he earned a first-class physics degree in 1925 and an MSc in 1926. After graduation, Randall worked for General Electric Company, where he did research involving luminescent powders used in discharge lamps. Progress in identifying fundamental mechanisms responsible for luminescence earned him a Royal Society fellowship to the University of Birmingham, where he worked in Professor

Marcus Oliphant's physics lab, starting in 1937. Oliphant's group at Birmingham included some fascinating characters, one of whom became so intertwined with Randall that the two cannot be separated: Maurice Wilkins (1916–2004).

SECOND IN COMMAND

Wilkins was born in New Zealand in 1916. When he was six years old, the family moved to Birmingham. Wilkins attended St. John's College, Cambridge, and earned his bachelor's degree in 1938. The degree was second-class, so he couldn't stay at Cambridge. He went home to Birmingham and approached Oliphant, who had been his instructor at Cambridge, telling Oliphant of his desire to pursue a research interest in luminescence. Since Randall had just

Maurice Wilkins (1916–2004). From Wikimedia Commons, user Materialscientist.

joined Oliphant's group with the same research interest, Wilkins was hired, and he worked for Randall on a luminescence project for his dissertation.

The anticipated war helped catalyze some very interesting developments at the University of Birmingham. Professor Oliphant was asked by the British Admiralty to develop a compact, powerful microwave transmitter for radar use. Randall and his colleague Harry Boot worked on the radar problem, leaving Wilkins and his dissertation project minimally supervised. That was fine with Wilkins, who got a chance to meet a few physicists who had been funneled to Birmingham by the war. They included Klaus Fuchs, Rudolph Peierls, and Otto Frisch. You may recall from chapter 12 that Fuchs turned out to become a Russian spy, and Peierls and Frisch (Lise Meitner's nephew) made the critical mass calculation that showed the feasibility of the uranium-235 bomb. Since these physicists were all foreign nationals, security didn't allow them to work on radar.

Randall and Boot quickly produced a functional transmitter, which was improved upon by General Electric engineers, and revolutionized radar systems used by both England and the United States during the war. Meanwhile, Wilkins plodded along on his thesis project and produced two significant papers; he also coauthored with Randall, although Wilkins pointed out that Randall did none of the work. The papers were sufficient to get Wilkins his PhD in 1940.

With the radar problem under control, Randall left to teach for a year at the Cavendish Laboratory in Cambridge, continuing his transition from industry toward his eventual goal of academic life. The rest of Oliphant's group, including Wilkins, Peierls, Frisch, and Fuchs, went to the United States to work on the Manhattan Project. Wilkins worked at Berkeley on the uranium isotope separation problem and Peierls, Frisch and Fuchs went to Los Alamos. While in Berkeley, Wilkins married an art student named Ruth, who became pregnant and then divorced Wilkins before the baby, a son, was born.

ACADEMICS FOR REAL

Back in Scotland, Randall was appointed Professor of Natural Philosophy at St. Andrew's. He wrote to Wilkins and offered him a job in biophysics research. The recently divorced Wilkins accepted and sailed back across the Atlantic, which was then freed of U-boats. Wilkins told Britain's *Encounter* radio program in 1999: "After the war I wondered what I would do, as I was very disgusted with the dropping of two bombs on civilian centres in Japan."[2] St. Andrew's was a bit out of the academic mainstream, and although the department thrived, Wilkins's personal research did not go well. Soon, he began exploring other avenues, principally around London and Cambridge. There were a few rows between Wilkins and Randall, but nothing serious. Fortunately, King's College London offered Randall the job of department head of physics and director of the Biophysics Research

Unit. Randall accepted, and Wilkins joined him. Randall's Lab (referred to as Randall's Circus) was built in a bomb crater in the college's quadrangle. Money flowed in from many sources as government committees recalled the pivotal role of new ideas in wartime—especially Randall's radar. Randall's relaxed management style and the infusion of plenty of research grant money made for a happy and productive group.

ANOTHER KEY PLAYER

Wilkins's initial research efforts involved ultrasonic probes of many organic molecules, including DNA, but they met with little success. One day Wilkins was approached by a physics graduate student of similar age, Francis Crick (1916–2004). Crick's bachelor's degree was in physics from University College London, but his PhD work was cut short by the war. His dissertation topic was the viscosity of water at high temperatures, which he labeled "the dullest problem imaginable."[3]

Francis Crick (1916–2004). Courtesy of Materialscientist, from Wikimedia Commons.

During the Battle of Britain, a bomb fell through the laboratory roof and destroyed Crick's experimental apparatus. Crick worked on underwater mines that could escape detection by minesweepers during the war. After the war, Crick became interested in applying physics to living systems and worked to determine the physical properties of cytoplasm at Cambridge's Strangeways Research Laboratories. Crick approached Wilkins, looking for a job in Randall's group. Wilkins took an immediate liking to him, regarding him as bright and energetic. The two became personal friends. When Wilkins told Crick of his personal research interests, Crick suggested Wilkins concentrate on proteins, not DNA. In spite of Wilkins's personal recommendation, Randall decided not to hire Crick, saying he was too boisterous and brash. Soon after, Crick joined the Cavendish Laboratory and explored the structure of proteins using X-ray diffraction.

DNA'S SIGNIFICANCE UNCOVERED

One of the King's College researchers, Geoffrey Brown, brought early news of key experiments that showed that DNA was the carrier of genetic information, so Wilkins's microscope research efforts were reduced so he could work more on DNA. Although he wasn't familiar with the techniques of X-ray crystallography, Wilkins was able to obtain very high-quality DNA samples for their research from Professor Rudolf Signer of Switzerland, a London conference participant. The only X-ray equipment the lab owned was designed to study relatively large single crystals, and so it was a bit large for the tiny DNA fibers. Nevertheless, research student Raymond Gosling (1926–2015) mastered the equipment, while Wilkins was able to prepare strands of the DNA on a tungsten wire frame. Randall suggested that air in the camera be replaced by hydrogen to sharpen the images, and part of a condom was used to seal the camera's hydrogen contents. The apparatus seemed cobbled together, but the results were the sharpest X-ray diffraction pictures yet. Not long after, the war surplus X-ray tube burned out, but the Chemistry Department offered the use of their large Raymax equipment so the work could continue. Some of the patterns obtained by Gosling implied that DNA might be wound into a helical (spiral) form. Reaching the limit of their equipment, they ordered a new fine-focus X-ray tube, along with a special micro-camera.

THE NEXT KEY PLAYER

To improve the lab's X-ray capability, in late 1950, an experienced X-ray researcher was added to the staff: Rosalind Franklin (1920–1958).

Franklin showed exceptional scholastic abilities from an early age and excelled at science, Latin, and sports. She went to Newnham College, Cambridge, and was awarded a fellowship to study physical chemistry at the University of Cambridge. She worked on classification of coals for fuels and gas masks, and earned her PhD in 1945. Next, she worked in Paris, applying X-ray crystallography to amorphous substances. Three years later, she accepted a fellowship to join Randall's group at King's College, London. Her original assignment was to work on X-ray diffraction from proteins and lipids in solution, but this initial assignment was redirected before it started.

Rosalind Franklin (1920–1958). From BioMed Central blogs, *On Biology*, July 25, 2013.

Here is a portion of the letter from Randall to Franklin:

4th December, 1950

Dear Dr. Franklin,

I am sorry I have taken so long to reply to your letter of November 24th. The real difficulty has been that the X-ray work here is in a somewhat fluid state and the slant on the research has changed rather since you were last here.

After very careful consideration and discussion with the senior people concerned, it now seems that it would be a good deal more important for you to investigate the structure of certain biological fibres in which we are interested, both by low and high angle diffraction, rather than to continue with the original project of work on solutions as the major one.

Dr. Stokes, as I have long inferred, really wishes to concern himself almost entirely with theoretical problems in the future and these will not necessarily be confined to X-ray optics. It will probably involve micros-copy in general. This means that as far as the experimental X-ray effort is concerned there will be at the moment only yourself and Gosling, together with the temporary assistance of a graduate from Syracuse, Mrs. Heller. Gosling, working in conjunction with Wilkins, has already found that fibres of desoxyribose [*sic*] nucleic acid derived from material pro-vided by Professor Signer of Bern gives remarkably good fibre diagrams.

CONFUSION BEGINS

This letter started the ball rolling on a misunderstanding that led to some sig-nificant consequences. Here's the next step in this dysfunction: When Rosa-lind Franklin began working at King's College, London, the first staff meeting wasn't attended by Wilkins, who was on holiday with a very attractive au pair who worked for some artist friends. Initial communication between Wilkins and Franklin was zero. For the next few months, Franklin cleaned up a few loose ends from her Paris assignment, then worked on setting up and fine-tuning the new X-ray apparatus and micro-camera. Franklin/Wilkins com-munication: still nothing substantive about DNA. Then, Wilkins made a pre-sentation at Cambridge to protein X-ray workers, including Francis Crick.

The talk included conjecture of the possible helical shape of the DNA molecule, its size, and its possible significance.

As Wilkins left the building, the first significant communica-tion between him and Franklin occurred. Franklin told Wilkins in no uncertain terms that he should stop doing X-ray work on DNA. She said he should "go back to your microscopes!" In fairness to both Franklin and Wilkins, there was one phrase in Randall's letter

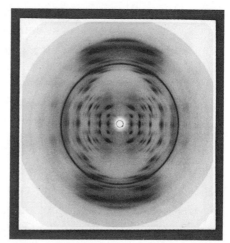

Wilkins DNA slide. Used with permission of King's College London © Raymond Gosling.

to consider: "as far as the experimental X-ray effort is concerned there will be at the moment only yourself and Gosling." Wilkins, on the other hand, had never seen the letter, nor had the issue been discussed with Randall.

Prior to this shocker, Wilkins had cemented himself firmly in DNA research by making interesting contacts with participants in DNA's past and future. Randall had been scheduled to speak at a conference held at the Zoological Station in Naples. The press of work was too great, so he sent his second-in-command, Wilkins, to summarize the activity at King's. Wilkins's talk was well received. He showed a slide that demonstrated clear X-ray diffraction of the DNA molecule in a photo taken by Raymond Gosling under Wilkins's direction.

In attendance was William Astbury, the pioneer in X-ray diffraction studies of DNA. Astbury had produced a similar photograph in 1939, but he acknowledged that Wilkins's results were much clearer. Also present was an extremely young fellow from America who was very excited about the potential crystalline structure of DNA as evidenced by the X-ray diffraction photograph. His name: James D. Watson.

THE LAST KEY PLAYER

James Dewey Watson was born in Chicago in 1928. As a youth, he never tired of asking why and spent hours poring over the *World Almanac*. He attended public schools and once won $100 on the TV program *Quiz Kids*. He was knocked out by a Jewish girl, Ruth Duskin (Feldman) on a Bible question.

Watson used the money to buy binoculars to continue the bird-watching hobby he shared with his father. In his own words, he was "far from the child genius,"[4] but he entered the University of Chicago at age fifteen because Robert Hutchins, president of the University of Chicago, believed in "getting the kids into college two years earlier."[5]

His first two years were not all that successful grade-wise, but Watson says they taught him three valuable things: read original sources, organize facts into rational schemes, and learn to think as opposed to relying on memorization. Dreams of ornithology took a backseat to genetics when

unidentified, James Watson, and Ruth Duskin (Feldman) in 1942. Reprinted from
Whatever Happened to the Quiz Kids? Perils and Profits of Growing Up Gifted,
by Ruth Duskin Feldman.

Watson read Erwin Schrödinger's book *What Is Life?* in his senior year.
By 1950, Watson had earned a PhD in zoology from Indiana University
(at age twenty-two) and was hot on the trail of the molecule that controls
genes: DNA. Postdoctoral work in the laboratory of biochemist Herman
Kalckar focused too much on chemistry to satisfy Watson's DNA interests.
But then came the meeting in Naples where Wilkins's X-ray slide captured
Watson's imagination.

THE TEAMS ARE IN PLACE

The Medical Research Council Unit for the Study of Structure of Biological Systems got a new member in September 1951: James D. Watson.
Located in the Cavendish Lab at Cambridge, the small group was headed

by Max Perutz, a chemist who had been collecting X-ray diffraction data on hemoglobin molecules for ten years. Head of the entire Cavendish Lab was Sir Lawrence Bragg, a Nobel Prize–winner in Physics for being the pioneer of X-ray crystallography. Bragg still participated in the research, trying to interpret the experimental results in terms of molecular structure. The unit also included John Kendrew and Hugh Huxley, who mostly worked on muscle cells. Rounding out the unit was Francis Crick, whom we met earlier in this chapter.

Recall that Crick had been a physicist before and during the war, but he changed to biophysics, possibly influenced by his reading of Schrödinger's book. At the Cavendish, Crick's thesis topic was protein crystallography, but his wide interests and ebullient personality carried him far afield. Although he was pleasant and polite, he visited many different labs and offered his opinions about theory at high volume to anyone who was interested. His explosive laughter made it impossible for him to hide, not that he was wont to do so. Crick and Watson shared an office at Cavendish, and Watson tried his best to interest Crick in his DNA quest. Crick had worked on protein structure for two years and wasn't prepared to jump ship quickly. Besides, DNA structure was being studied in another lab at King's College London by the disjoint team of Wilkins and Franklin. Complicating the competitive picture between the Cavendish and Randall King's group was the personal friendship between Crick and Wilkins.

FIRST SCORE: KING'S TEAM

Working away on X-ray diffraction, Franklin soon discovered that there are two distinct forms of DNA. One form was labeled "A," in which the molecules are less hydrated and shorter, while the other, more heavily hydrated form consisted of longer molecules and was labeled "B." Wilkins took the opportunity to suggest to Franklin that they work together, but Franklin angrily refused. Randall imposed a settlement by assigning research on the A form to Franklin using the excellent Signer DNA samples, while Wilkins would work with the B form using samples obtained from biochemist Erwin Chargaff at Columbia University. The Chargaff samples turned out to be too degraded for effective X-ray studies, so Wilkins made little progress.

CAVENDISH TEAM'S TURN

Meanwhile, at the Cavendish, Watson and Crick decided to take a new approach. Although neither one was technically assigned to analyze DNA, their fascination drove them to take a page from a master's book. Earlier in 1951, American chemist Linus Pauling at Caltech had scooped everyone, but most irritatingly the Cavendish lab, by discovering the basic structure of proteins—the alpha helix. This was just the latest triumph for Pauling over Bragg, who was Pauling's chief competitor. Part of Pauling's solution technique was building physical models of the molecular components, then fitting them together.

In November of 1951, Crick and Watson built a physical model of the DNA molecule as a three-chain helix with phosphate groups on the outside. Part of the basis of the model was Watson's recollection of the water content of the molecule in a presentation by Franklin. Crick invited Wilkins to come to the Cavendish to see the model. Wilkins accepted and brought his collaborators Bill Seeds, Bruce Fraser, Franklin, and Gosling the following day. The visit was a disaster for Crick and Watson. Franklin pointed out that the water content was all wrong (faulty memory of Watson) and the molecule wasn't proven to be helical in her research. In her view, model building should only be attempted when the structure is well known, which it certainly was not.

BRAGG'S MORATORIUM

As a direct result of the meeting, Bragg, directed that Crick and Watson should give up their DNA work and leave the DNA analysis to King's. Crick and Watson showed evidence of their cooperation by sending copies of the metal jigs they had used for model building.

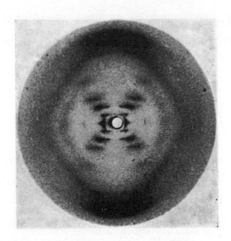

Photo 51. Used with permission of Kings College London, © Raymond Gosling.

All was quiet for most of 1952. Crick worked on his thesis topic, the structure of hemoglobin; Watson took up the topic of the tobacco mosaic virus (TMV), a convenient cover since it has a substantial DNA component; and the King's group continued their X-ray diffraction experiments. Franklin's work provided no confirmation of a helical structure for DNA. She sent Alex Stokes (mentioned above in Randall's letter to Franklin) and Wilkins a prank notification of the death of the DNA helix. Wilkins and Stokes took this joke somewhat more seriously than intended, however. In fact, Franklin could find no good evidence for helical structure in the A-form, but she never suggested the B-form was anything but helical. In fact, one of the photographs taken by Gosling in May of 1952, the now-famous Photo 51, showed very clearly the helical nature of the B-form of DNA.

CATALYSTS

In chemistry, a catalyst is a substance that assists a chemical reaction without becoming a permanent part of the products. In 1952, there were two human catalytic agents whose actions had a direct bearing on the determination of DNA structure. The first was Rosalind Franklin, when she chose to leave King's. In June, Franklin decided to spend the third year of her fellowship at Birkbeck in the lab of Professor J. D. Bernal. Randall was notified, asked Franklin to leave all her DNA work at King's and to not continue studying DNA at her new position. The second catalyst consisted of two young Americans who arrived in the late fall to work at Cavendish: Peter Pauling (oldest son of Linus Pauling) and Jerry Donohue. Donohue shared the office with Watson and Crick, and Pauling spent a lot of time with the group. These catalysts combined to help win the DNA structure race for Bragg's Cavendish laboratory.

Hang on, the slow pace of normal research is now accelerated as rapid progress was then made, culminating in the determination of the structure of DNA.

- Christmas 1952: Linus Pauling announced in a letter to his son Peter that he has solved the structure of DNA. Peter passed this along to Watson and Crick, who were disheartened but pressed on regardless. They had agreed not to work on DNA, but they couldn't stop thinking about it.
- January 28, 1953: Rosalind Franklin, at her last seminar before leaving, summarized her work on the A-form of DNA but made no mention of a helix or anything else about the B-form.
- January 28, 1953: Two preprints of Pauling's paper arrived, one to Peter and one to Bragg. Watson grabbed Peter Pauling's copy and scanned it hurriedly. He was overjoyed to note that Pauling's model was quite similar to the Watson/Crick model from 1951—and therefore, wrong. On the other hand, Pauling was on the trail, and it probably wouldn't take him long to correct his errors.
- January 30, 1953: Watson rushed to King's and offered to show Pauling's paper to Franklin. In a scene judged to be confrontational by Watson, Franklin implied that she was busy and not interested in more faulty DNA structure arguments. Watson beat a hasty retreat and ran into Wilkins. Wilkins was interested in Pauling's error, and showed Watson Photo 51, which had been given to him by Gosling. Instantly, the strong X-pattern revealed to Watson the helical nature of the DNA molecule.
- February 1953: Cavendish lab director Bragg turned Watson and Crick loose to allow them to return to building a DNA model. He was not about to lose another competition to Pauling. Wilkins also agreed that Watson and Crick could go back to working on DNA. As their model building ran into snags, Jerry Donohue suggested

Watson and Crick with DNA model. © A. Barrington Brown / Science Source.

they use a structural variation on their molecular bases. It worked beautifully.

- Early March: After a bit of fiddling, the Watson/Crick model of DNA was complete. They invited many to see it, and it withstood all criticisms. The full paper is available online.[6]

In 1962, the Nobel Prize in Physiology or Medicine was awarded to Crick, Watson, and Wilkins "for their discoveries concerning the molecular structure of nucleic acids and its significance for information transfer in living material."[7] The chemistry prize was given to Perutz and Kendrew, "for their studies of the structures of globular proteins."[8] Linus Pauling wasn't shut out completely. He won the 1962 Nobel Peace Prize.

Bonus Material: Franklin/Wilkins Internet interview. See To Dig Deeper for details.

AFTERMATH

The exclusion of proper Nobel credit for the critical work of Rosalind Franklin created a controversy in the mid-1960s, but Franklin's death from ovarian cancer in 1958 precluded any real Nobel credit, since the prize is awarded only to living scientists. Her work at Birkbeck on the tobacco mosaic virus was top-notch, and she amassed well-deserved credit. Franklin remained scientific and personal friends with Francis Crick and his wife, Odile, until her death.

In his 1968 book *The Double Helix*, James Watson tends to caricature Franklin, referring to her as "Rosy," and "Wilkins' assistant."[9] In response, a book written by Anne Sayre (*Rosalind Franklin and DNA* [New York: W. W. Norton, 2000]), a personal friend of Franklin's, set the record straight, pointing out that "Rosy" was an aunt in the Franklin family, not a preferred nickname for Rosalind, and that Franklin was hired as a far more experienced X-ray analyst than Wilkins, and not just an assistant.[10]

Some think the pendulum has swung too far, and Franklin, as her younger sister Jenifer Glynn puts it, "has become 'the forgotten heroine.' Her story has been adopted by feminists as a symbol of a woman struggling and unacknowledged in a man's world. This would, I think, have embarrassed her almost as much as Watson's account would have upset her. It suited the feminism of the 1960s and 1970s to portray her as a victim of male dominance, but she would have thought of herself simply as a scientist whose achievements should have been judged on their own terms."[11]

KUDOS

John Turton Randall was knighted in 1962 and retired to Edinburgh in 1970.

Maurice Wilkins became a Fellow of the Royal Society in 1959, the same year he married Patricia Chidgey. Wilkins remained at King's College London until he retired in 1981. He became president of the British Society for Social Responsibility in Science in the late sixties and continued in that position for twenty years. In 2003, he released his autobiography, *The Third Man of the Double Helix*.

Used with permission from Sidney Harris.

Raymond Gosling was a fellow in an awkward position: supervised by two people who didn't get along and dependent on them to get his PhD. Fortunately, he had a good sense of humor. He called Randall "JT" or "the old Man" or "King John." When he was assigned to Franklin, Gosling said he was "the slave boy handed over in chains." He called Wilkins "uncle Maurice."[12] Since he was involved in taking so many critical X-ray diffraction pictures, Gosling might have written a book titled *The Fourth Man of DNA*, but he didn't. He received his degree in 1954 and lectured in physics at St. Andrew's and the University of the West Indies. He then became a lecturer and reader at Guy's Hospital Medical School, London.

Francis Crick completed his dissertation and received his PhD in 1954. He did his postdoctoral work at Brooklyn Polytechnic, then went back to Cambridge until 1976. Later, he went to La Jolla, California, to the Salk Institute and worked on neuroscience, eventually studying consciousness.

James Watson taught biology at Harvard from 1956 to 1976. In 1968, he wrote the book *The Double Helix*. In the preface, he notes: "I have attempted to re-create my first impressions of the relevant events and personalities rather than present an assessment which takes into account the many facts I have learned since the structure was found."[13] The book delivers on this promise and generated much controversy in the process. Some of the "Rosy" comments are detailed above, but Francis Crick was none too happy with the book either. Watson seems comfortable with controversy, self-generated or not. Fellow Harvard professor Edmund O. Wilson called Watson "the most unpleasant human being I had ever met."[14]

Watson became director of Cold Spring Harbor Laboratory, New York, in 1968 and continued in that position until 2007. In that same year, he was quoted in the *Times* of London as saying "[I am] inherently gloomy about the prospect of Africa [because] all our social policies are based on the fact that their intelligence is the same as ours—whereas all the testing says not really."[15] He went on to say that despite the desire that all human beings should be equal, "People who have to deal with black employees find this not true."[16] The furor this quote generated led the trustees of Cold Spring Harbor to ask for his retirement. He is now Director Emeritus of Cold Spring Harbor Laboratory.

And as for auctioning the Nobel medal, it was bought by Alisher Usmanov, a major shareholder in the Arsenal Football Club and said to be the richest man in Britain. In a ceremony, the medal was returned to Watson, and the tycoon said he was "distressed" that Watson felt forced to sell it.[17]

So, what was the motivation for the sale? Several quotes from Watson have a bearing here:

"No-one really wants to admit I exist."

"I want to re-enter public life."

"I really would love to own a [painting by David] Hockney."[18]

Another possible influence might relate to the fact that Francis Crick's seven-page handwritten letter to his son explaining the DNA discovery had just sold in 2013 for six million dollars, an all-time record for the auction of a letter.

Oh, one last thing. Watson served as head of the Human Genome Project from 1988 to 1993. That project and its aftermath are the subject of the next chapter.

CHAPTER 14

J. CRAIG VENTER, JAMES WATSON, AND MICHAEL HUNKAPILLER RACE FOR THE HUMAN GENOME

Used with permission from Sidney Harris.

W hat in the world is a genome? In 1920, Hans Winkler, botany professor at the University of Hamburg, Germany, combined the terms *gene* and *chromosome* into one, calling them jointly the genome. The study of genetics in biology percolated right along. Gregor Mendel's work from the 1800s was rediscovered in 1900. It was clear that an organism's traits depended on the prior generation's characteristics through what were called genes and chromosomes.

But what are genes and chromosomes, and by what mechanisms are traits passed to succeeding generations? The answers to these questions came into sharp focus just after World War II, when biology experienced a series of visitations from other fields. They weren't quite strong enough to be called invasions, but their effect on biology was stronger than mere interactions. A related example of this phenomenon would be the reaction of the field of geology to meteorologist/climatologist Alfred Wegener's theory of continental drift, which we saw in chapter 8.

Visitation 1 was accomplished by physicists. That was what chapter 13 was all about. Physics-educated Crick, Wilkins, Franklin, and Randall identified the structure of DNA using X-ray crystallography techniques borrowed from physics. (James Watson was the only participant with degrees in biology.)

Another set of visitors were the biochemists. (Actually, both biophysics and biochemistry had existed prior to the DNA discovery, but their popularity and credibility soared enormously after DNA became a hot topic.) Even before the structure of DNA was worked out, biochemists had learned that most of the workings of living organisms are accomplished by proteins. With the discovery of DNA, it became clear that the organism's plan was contained in its DNA, then transferred to its RNA (ribonucleic acid), which, unlike DNA, ventures out of the cell's nucleus and helps build the necessary proteins. So, a chromosome is a collection of genes, and each gene is a segment of DNA.

DECIPHERING THE GENOME

And now we can finally see just what constitutes a genome. An organism's genome is its complete set of DNA, including all its chromosomes, which in turn contains all its genes. That seems like a large package of information, and it is. The genome contains all the information necessary to build and allow that organism to function and includes input for the organism's next generation. To compare the genomes of different

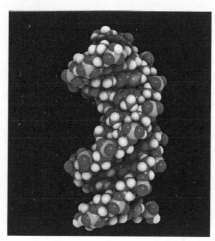

Model of DNA molecule. From Wikimedia Commons, user Spiffistan.

organisms, we need to delve into the innards of DNA.

DNA's two helices are joined by what are called nucleotide bases. The helices are linked by only four bases: cytosine (C), guanine (G), adenine (A), and thymine (T). These bases join together only in certain pairings: C with G, and A with T. These base pairs make up the backbone of the DNA molecule. The details of how the DNA molecule functions are truly fascinating. You are encouraged to consult the To Dig Deeper section in the back matter if you want more information.

Now we are in a position to compare the genomes of different organisms, in terms of the number of base pairs in the genome.

Organism	Number of base pairs in the genome	Comment
Mycoplasma genitalium	580,000	Smallest true organism
E. coli K-12	4,639,000	bacterium often studied in labs
S. cerevisiae	12,500,000	brewer's yeast
D. melanogaster	123,000,000	fruit fly
Mus musculus	2.8 Billion	mouse
Homo sapiens	3.3 billion	human
Picea abies	19.2 billion	Norway spruce

To put these numbers of base pairs into perspective, if you counted one base pair per second, it would take you over one hundred years to enumerate the human genome.

In the 1970s, clever methods were developed to read the DNA sequence, but they were not very fast and were quite expensive. A biology graduate student might spend several years to sequence a several-thousand-base-pair portion of a genome. Biology's instruments improved, however, and by the mid-1980s, the lure of sequencing the entire human genome was enormous.

> The total human sequence is the grail of human genetics . . . an incomparable tool for the investigation of every aspect of human function.
> —Walter Gilbert (1980 Nobel Laureate)[1]

Seeking the grail conjures up the image of many knights, each following their own arduous path, seeking an extremely significant object. True. The

Used with permission from Sidney Harris.

human genome took all that and more. It required a whole new set of visitors to biology—instrument makers, venture capitalists, information technologists, politicians, Big Science executives, and one talented but extremely impatient fellow: J. Craig Venter. Let's start with him.

John Craig Venter was born in 1946 in Salt Lake City, Utah. Shortly thereafter, the family moved to California, where Craig loved boats and surfing, but not school, where he earned lots of Cs and Ds. Although he opposed the Vietnam War, he wound up as a corpsman in a field hospital's intensive care unit. Perhaps moti-

J. Craig Venter (1946–). From Wikimedia Commons, user Calliopejen.

vated by the wounded, maimed, and dying marines, he decided to study medicine when he got out. He began his studies at the College of San Mateo community college in California, then transferred to the University of California, San Diego, where he earned a BS in biochemistry in 1972 and a PhD in physiology and pharmacology in 1975.

Venter's first academic position was at the State University of New York at Buffalo, where his major research interest was in the protein that acted as the brain's receptor for adrenaline. Eventually, it became clear to him that he didn't have access to enough of the tools of molecular biology to make the progress he wanted, progress that included finding the DNA sequence of the particular protein he was researching. Even with the collaboration of Caltech's biology professor Leroy (Lee) Hood and his postdoctoral student Michael Hunkapiller, progress was too slow for Venter.

OUT OF THE FRYING PAN . . .

In late 1983, the answer sought Venter. He was recruited by the National Institutes of Health (NIH) in Bethesda, Maryland, and became chief of the Receptor Biochemistry and Molecular Biology Section, National Institute of Neurological Disorders and Stroke (NINDS). After about a year of using traditional methods, Venter's team had sequenced the genetic code for the first human brain receptor protein. Its results were written up in a paper published in a series of letters sponsored by the Federation of the Societies of Biochemistry and Molecular Biology (FEBS), but Venter was frustrated by the slow pace of the sequencing.[2] To his surprise, there was an article in a recent issue of *Nature* by Lee Hood's group at Caltech about a new machine that performed automated DNA sequencing. Michael Hunkapiller had joined a biotech company named Applied Biosystems (ABI), which was about to market their brand-new sequencer. Venter ordered one of their first machines, the ABI 373A. It was delivered in February 1987 and installed in his own office where he helped work out the bugs. A sixteen-hour run of the new machine analyzed a sample that would have taken a week the old way. Venter was undoubtedly impressed by the speed, but there was still a long way to go to make a complete analysis of adrenaline.

THE LURE OF THE HUMAN GENOME

Venter paid attention to the discussion that swirled through the biotech world about human genome mapping. He read the arguments about whether a human genome mapping project should be government-funded, pursued at research institutions, or investigated by private industry, and whether it should be researched by individual countries or coordinated by an international consortium. Although some weighed in with the notion that it would require all biology graduate students' efforts for the next fifteen years and cost three billion dollars, Venter thought it would be a worthwhile effort. Venter convinced his boss, Ernst Freese, and his boss, Irwin Kopin, that he had something to contribute. Part of Venter's lab was converted to the NINDS DNA-sequencing facility. With the rapid purchase of three additional sequencers, Venter's lab became the largest DNA-sequencing facility in the world.

THE BIG TIME

In early 1988, the Ad Hoc Advisory Committee on Complex Genomes met and announced the appointment of James Watson (yes, *the* James Watson) as the associate director of the newly formed Office of Human Genome Research.[3] Watson had his plate full with guiding the setup of such a huge operation, but he must have been delighted to meet Venter, who provided a genome-sequencing capability in NIH's own backyard. At their first meeting, Venter told Watson he needed a few million dollars to sequence a single chromosome, the X. Watson told Venter that amount was insufficient, and to write up a proposal asking for five million dollars.

BUREAUCRACY TANGLES

Intense political pressure from several directions snarled Watson's resolve, however well intentioned. Venter's proposal eventually underwent four revisions over a period of two years and was never funded. The overall project became the Center for Human Genome Research[4] and officially began its operation in October 1990. It was set up to be a worldwide effort, with the majority of the work being done by various governmental facilities and universities in the United States, and about a third in the United Kingdom, France, Germany, and Japan.

IMPATIENCE, AGAIN

Venter was too impatient to wait for the federal bureaucracy to grind its way to funding his project. He got an idea about how to shortcut the process. To find the active genes in a particular cell type, he first extracted the RNA from the cell. Since RNA is built according to the plan contained in the DNA, it contains the nucleotide base pair sequence from active portions (genes) of the original DNA. The RNA could then be converted to more stable DNA (called complementary DNA, cDNA) and attached to a bacterial chromosome for storage, using the cut-and-paste techniques available through special molecules that severed DNA at known locations. These are called restriction enzymes. Complementary DNA was a standard resource in molecular biology labs all around the world, so its availability was ensured. Next, the cDNA would be sequenced and compared to other sequenced genes. This idea, called expressed sequence tags (EST), was not new to Venter. It was first published by Paul Schimmel, professor of chemical biology at Scripps Institute in 1983, and was used extensively by the renowned geneticist Sydney Brenner of the Laboratory of Molecular Biology at Cambridge and others in the late 1980s. But, thanks to Venter's ABI sequencer and workstations, no one had the sequencing capability of his lab.

In June 1991, Venter reported in *Science* that by sequencing ESTs he had identified about 330 genes active in the human brain. In one stroke, Venter had identified and sequenced more than 10 percent of the existing world total of known human genes—all in a matter of months. In his customary direct fashion, Venter pointed out that "improvements in DNA sequencing technologies have now made feasible essentially complete screening of the expressed gene complement of an organism."[5]

The immediate negative reaction of some biologists was further fueled by Venter's next paper, published in *Nature*. There, he reported another 2,375 human genes expressed in the brain—double the number of genes sequenced by the rest of the scientific community at the time. A major worry was that the cDNAs sequenced by the EST procedure of Venter would be funded as a cheaper alternative to sequencing the entire human genome. This approach would miss the subtleties of gene regulation, switching, and control because binding sites for activators and repressors would not be sequenced.

PERILS OF PATENTS

A cause of additional trouble was the patenting of ESTs. The NIH Office of Technology Transfer approached Venter, asking him his intention about patents on the large amount of gene-sequencing data his lab was generating. Venter's limited experience with patents when he researched at the University of Buffalo was that they delayed the publication of scientific information, and so he was definitely not thrilled. Intellectual property had become a hot-button issue in biotechnology, and the law was unclear on ESTs. The Office of Technology Transfer's idea was to apply for a patent, then sort out the details. Venter agreed, but on two conditions: first, the research results would be published anyway, and second, the decision to file would be brought to the attention of top management, especially Watson, to make sure this was the right approach. So, a patent application was filed for the first 330-plus genes prior to Venter's first publication in *Science*, and 2,421 more genes were added to the application before the *Nature* article was published—all in Venter's name. The furor arose quickly and never abated. French research minister, Hubert Curien, said that "a patent should not be granted for something that is part of our universal heritage."[6]

BLOWUP

Less than a month after the *Nature* article, there was a Senate hearing of the Committee on Energy and Natural Resources. Venter described his EST research and raised concerns about the patent efforts, which had not been publicly disclosed. The room was suddenly filled with shouts from Watson, who said it was "sheer lunacy" to file such patents and that he was "horrified" because "virtually any monkey" could use the EST method. Venter was shocked that Watson had held him responsible for what Watson later referred to as "Venter patents," when the idea had originated in the Technology Transfer Office.[7]

But a *Washington Post* reporter, Larry Thompson, identified his

Bernadine Healy (1944–2011), Senator Barbara Mikulski, and Dr. Vivian Pinn. From Wikimedia Commons, user Hildabast.

impression of the real combatants: Watson and NIH director Bernadine Healy.

Bernadine Healy (1944–2011) had distinguished educational credentials from Vassar College and Harvard Medical School and had completed her internship and residency in cardiology at Johns Hopkins University. Eventually, she was appointed by President Ronald Reagan to the position of deputy director of the White House Office of Science and Technology Policy. During this tenure, she was involved in a 1985 spat with Watson, in which he complained about genetic technologies regulations by claiming that in the White House, "the person in charge of biology is either a woman or unimportant. They had to put a woman someplace."[8]

Subsequently, Healy was director of the Research Institute at the Cleveland Clinic Foundation when President George H. W. Bush tapped her in 1991 to become director of the NIH, its first woman head. As NIH director, Healy was now Watson's boss. She believed the patent application to be appropriate and dismissed Watson's objections as "a tempest in a teapot." She instructed Watson not to criticize Venter in public and asked

Venter to consult with her on human genome research. Watson resigned in April 1992 (by fax), calling his position "untenable." Meanwhile, Venter had applied for ten million dollars to expand his sequencing operation. His proposal was vigorously rejected by NIH peer review.[9]

NEXT VISITORS: VENTURE CAPITALISTS

Wallace Steinberg, head of the HealthCare Investment Corporation and inventor of the Reach toothbrush, put up seventy million dollars to lure Venter into the private sector. This venture was called The Institute for Genomic Research (TIGR). Venter bit. He resigned from NIH in July 1992. Starting with thirty ABI 373A Sequencers, seventeen ABI Catalyst workstations and a Sun SPARCenter 2000 computer with relational database software, Venter set out to increase EST-sequencing production and to sequence genes from model organisms as well. At a cost of one hundred thousand dollars per machine, deep pockets were needed to fund the venture, but Venter, once funded, would have free rein to pursue his sequencing ideas. Human Genome Sciences (HGS) was set up as a sister company to explore commercial aspects of genomic research. Venter was delighted. He proclaimed, "It's every scientist's dream to have a benefactor invest in their ideas, dreams and capabilities."[10] The only catch was that HGS would have only six to twelve months to review Venter's data before publication. His scientific colleagues were decidedly less enthusiastic. Some even referred to him as "Darth Venter."[11]

Meanwhile, back at the public consortium, a new director for the National Center for Human Genome Research was announced. The well-respected University of Michigan medical geneticist Francis Collins became the center's second director. As work continued, the consortium posted some impressive results. In 1996, the complete genome of brewer's yeast (*Saccharomyces cerevisiae*) was completed. This single-celled organism contains six thousand genes constructed out of twelve million nucleotide base pairs in its DNA. Over a hundred laboratories, located in Europe, the United States, Canada, and Japan completed the sequence.

Yet, as the halfway point of the scheduled time for the Human Genome Project approached, less than 3 percent of the genome was sequenced and the public consortium costs were running way over budget. Francis Collins appealed for more speed and more novel and productive ideas, but only slight progress resulted.

SHOTGUN SEQUENCING

While the public consortium tried to speed up its progress, Venter's lab, TIGR, tried a totally new tactic: shotgun sequencing. Johns Hopkins University researcher Hamilton Smith, who had discovered restriction enzymes almost twenty years earlier, had a radical idea. First, shear the DNA into thousands of random-sized pieces using sound waves, then sequence the pieces individually using the ABI machines. Store all the sequence data in a computer and let specially written software find overlaps so the pieces could be stitched together on the mathematical basis of pattern recognition to form one contiguous DNA. The technique seemed to work in simulations, and J. Craig Venter didn't shy away from the gamble. TIGR sequenced the entire genome of the *Haemophilus influenzae* bacteria within thirteen months, for less than half the Human Genome Project sequencing cost. In short order, TIGR then completed the sequence of *Mycoplasma genitalium*, the smallest free-living organism known, as well as several simpler genomes. Venter's reputation soared among his fellow researchers as more valuable sequencing information was made available for study.[12]

The shotgun sequencing technique worked—for bacteria—but still wasn't fast enough for the Human Genome Project to finish in time. That was soon to change. Late in 1997, the relationship between Venter's TIGR and its sister company Human Genome Sciences unraveled completely. Although HGS still owed TIGR thirty-eight million dollars, Venter released them from the obligation. This action bought Venter the freedom to release sequencing information faster, without a delay for HGS review.[13]

But Venter had even bigger plans, which revolved around the talents of Mike Hunkapiller. Since developing the original sequencing machine, the ABI 373A, with Leroy Hood in the late 1980s, Hunkapiller had not only made a number of improvements; he had instituted a substantially changed process. The earlier technique involved running DNA fragments down lanes through a gel for a portion of the separation. Now, Hunkapiller had developed a method in which the DNA was sent down thin, liquid-filled capillary tubes. With many more lanes available in a single run as well as other speed-enhancing improvements, the new machine, the ABI PRISM 3700, was about eight times faster than existing machines. After showing Venter the prototype, Hunkapiller popped the question: Would Venter team up with him to sequence the entire human genome? After some initial skepticism, Venter agreed. Something new had to be done, because the techniques that worked so well on bacteria couldn't be applied directly to the thousandfold-larger human genome.

Venter relished the challenge. After some initial consultation—more like a warning—with Human Genome Project director Francis Collins, Venter announced the formation of his new company. Its main goal: sequence the entire human genome and accomplish it within three years, substantially sooner the HGP schedule. His new company's name: Celera, from the Latin *celeris*, meaning "swift." The company's motto: "Speed matters. Discovery can't wait."[14]

Venter had done it again. The scientific world was sent spinning, but this time Venter's solid record of accomplishment made the critics much more circumspect. Maybe he *could* do it. After all, from Venter's standpoint, it was quite a risk. He had a barely tested prototype sequencing machine and no computer software because the old methods wouldn't work on the new genome. For his next move, Venter opened the door to biology's newest visitor: computer programmers, who Venter called algorithm scientists. Like the other visitors discussed earlier, some computer programmers became biology's permanent house guests and were rewarded with a new name for their field: bioinformatics. Stitching together overlapping sequences of nucleotide base pairs to create a whole genome was a significant computing problem, but Venter's massive investment in high-end computing equipment and expertise paid off. His team wrote a program that seemed to work.

As a test, Venter sequenced biology's favorite model organism, *Drosophila melanogaster*, the fruit fly. Both the machines and algorithms functioned well. The 165 million nucleotide base pair, 13,600 gene DNA, was sequenced in less than four months, just in time to burn it into CD-ROMs that graced every seat at a scientific meeting held the day before the release of the genome paper in *Science*.[15]

The Human Genome Project wasn't sitting idle through all of Venter's maneuvering. With increased funding from many sources, especially the Wellcome Trust in the United Kingdom, the consortium bought new sequencing machines (some from ABI, and some from Michael Hunkapiller's competitors) and stepped up their efforts, revising the timetables accordingly. If a race was on, so be it.

Although the competing parties spoke periodically, tensions were inflamed mercilessly by the communications media, especially given Venter's direct manner and Collins's personable but firm style. As the end of the "race" neared, news of the rough edges of the two groups' relationship reached all the way to the White House. President Bill Clinton told his science adviser, Neal Lane, to "fix it . . . make those guys work together."[16]

J. Craig Venter, President Clinton (1946–), and Francis Collins (1950–).
Courtesy AP/Worldwide Photos.

The job fell to Ari Patrinos, the Department of Energy's genome director. Patrinos had known both Venter and Collins socially, so in May 2000, he invited both of them to his Rockville, Maryland, townhouse for pizza and beer. They came, perhaps grudgingly, but reached an agreement to work together. The completion of the genome sequence was announced on June 26, 2000, the only day available on President Clinton's calendar. In a satellite linkup with UK prime minister Tony Blair, President Clinton said, "Modern science has confirmed what we first learned from ancient faiths. The most important fact of life on this earth is our common humanity."[17]

In the press conference that followed, Venter apologized for the absence of Mike Hunkapiller but explained he had contracted chicken pox. Venter said that if Hunkapiller attended, he would have to sit on the Public Consortium side.

THE RACE CONTINUES

Despite the media hype, the race to sequence the human genome was actually a race to a new starting line. To restate the problem: DNA has the plan for an organism's complete function. But before the function can be carried out, the plan must be transcribed into RNA, which in turn is translated into proteins, which then go on to build the cell's structure and help carry out its functions.

Proteins are the molecules that actually do the work of sustaining life. The genome tells RNA what proteins to build, but variations occur (proteins fold, interact, get sugars or methyl groups attached, etc.) before they actually carry out their multiple missions, eventually producing traits. Now you can see why referring to the human genome as the grail may be too simplistic. The complete collection of proteins, called the proteome, would be a logical but extremely difficult next step. Roy Whitfield, CEO of Incyte, one of the leading biotech companies said, "I would describe it as the beginning of thousands of races. If you have colon cancer, the race is about curing colon cancer. If you have arthritis, it's a race to cure arthritis. It's the start of a really long race to have a tremendous impact on human health."[18]

CHAPTER 15

TEN HONORABLE MENTION MINI-CHAPTERS

T here are many other people who had an impact on science and deserve recognition. The following mini-chapters feature those who merit honorable mention.

15.1 WILHELM CONRAD RÖNTGEN

I have discovered something interesting but I do not know whether or not my observations are correct.

—Wilhelm Conrad Röntgen[1]

Wilhelm Röntgen (1845–1923). This file comes from Wellcome Images, a website operated by Wellcome Trust, a global charitable foundation based in the United Kingdom. From Wikimedia Commons, user Fæ.

Röntgen was born in 1845, in Lennep, Germany. Shortly thereafter, he moved with his family to Apeldoorn in the Netherlands. In his youth, he enjoyed making mechanical gadgets and hiking outdoors, but he wasn't particularly good at academics.

In 1862, he was expelled from a technical school in Utrecht because he refused to tell which classmate had drawn a caricature of a particularly unpopular teacher. He tried attending classes at the University of Utrecht but was denied credit since he had no high school diploma. Hearing that one could enter the ETH Zurich

(Swiss Federal Polytechnic) without a high school diploma as long as they could pass the entrance examination (recall Einstein's experience in chapter 9?), Röntgen journeyed to Zurich, but he arrived two days late for the exam. An eye infection had slowed him down, but ETH officials allowed him to take the exam anyway. He passed and was admitted. He studied with Professors August Kundt and Rudolf Clausius, and earned his PhD in 1869.

Röntgen married Anna Bertha Ludwig of Zurich, and then embarked on an academic career at several universities in Germany. In 1895, Röntgen taught at the University of Wurzburg and studied the physics of passing electrical currents through extremely thin gases at low pressure. Others who studied similar phenomena were Eugen Goldstein, Johann Hittorf, William Crookes, Heinrich Hertz and Phillip Lenard. Generically, the tubes used were called Crookes tubes, and the observed rays were called cathode rays. These tubes are similar to those at the heart of early TV sets, and are called cathode-ray tubes or CRTs. Many other physicists were experimenting with the same tubes, but Röntgen noticed something unusual that the others missed. Even though the normal cathode rays (actually electrons) were blocked, fluorescent material would light up as far as two meters away from the tube. This implied that there was another beam besides the cathode rays, but this one was invisible. The beam also fogged photographic film. If objects of variable density were inserted in the beam, images of the objects' interiors appeared on the film. Röntgen had been closeted in his lab for weeks working on this phenomenon. He emerged to demonstrate it to his wife, Anna. He had her insert her hand in the beam, and this is the picture he got.

Anna's comment was: "I have

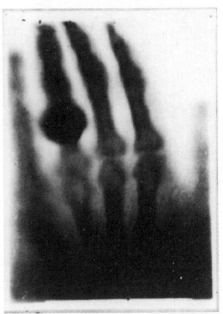

X-ray of Anna's hand. From Wikimedia Commons, user Melamed katz.

seen my death."[2] In fact, she lived another twenty-four years and died of intestinal cancer.

Since he knew so little about these rays, Röntgen called them X-rays, X representing the traditional mathematical symbol for an unknown. Although X-rays were almost immediately used in medical diagnostic practice, Röntgen refused to patent them, saying they were for the good of humanity. He also resisted those trying to name them "Röntgen rays," saying he preferred the X-ray name.

An extremely curious connection exists between Röntgen and Einstein. They had the same self-proclaimed enemy, Phillip Lenard. We saw in chapter 9 how Lenard attacked Einstein publicly in 1923, but it turns out that Lenard attacked Röntgen as well. Röntgen had won the first Nobel Prize in Physics in 1901 for his work with X-rays, and Lenard had won his Nobel in 1905 for his work with cathode rays. Both used similar tubes, and Lenard had actually supplied Röntgen with one of the tubes he used for his pioneering work. Lenard claimed priority in discovering X-rays, as he (and several others) had undoubtedly observed some aspect of them but had neglected them in their major focus on cathode rays.

Röntgen died in 1923. He enjoyed work and long walks in the country almost to the end.

15.2 PERLMUTTER, RIESS, AND SCHMIDT DISCOVER THAT THE EXPANSION OF THE UNIVERSE IS ACCELERATING

Accelerating? The universe is expanding faster and faster? How did we miss this until 1998? Maybe we should blame Einstein. Or, better still, Einstein and Willem de Sitter.

Prior to the development of Einstein's general theory of relativity, physics and astronomy had pretty much charted their own courses. Astronomy was busily trying to find out what was out there, while physics pretty much left it alone. Well, there was that relationship between the work of Newton and Halley (recall chapter 3), but the real kicker came when the ink was barely dry on Einstein's general relativity and de Sitter applied the theory to the entire universe. The result: As Einstein's equations stood,

Used with permission from Sidney Harris.

for solutions to be stable, the universe had to be either expanding or contracting. But common wisdom said that the universe wasn't doing either of these things. The universe seemed perfectly stable. The fixed stars were, well, fixed. As you may recall from chapter 10, Einstein added a term to his equations to make them describe a stable universe. It was called the cosmological constant. It didn't please Einstein much, but it seemed necessary at the time.

Hubble certainly can't be blamed. His 1929 observations using the hundred-inch telescope (with Milt Humason; see chapter 11) showed that the universe was expanding. Einstein was delighted, since he was now able to remove that pesky cosmological constant that he regarded as a blunder. But was the universe's expansion speeding up (acceleration), retaining the same velocity, or slowing down (deceleration)? Hubble's distance measurements weren't precise enough to reveal that information. Recall that Hubble's method of finding the distance to faraway galaxies was based on

Cepheid variable stars (recall Henrietta Swann Leavitt's period-luminosity relation from chapter 11). Cepheid variables couldn't be found in distant galaxies because they were too dim. So, it would seem necessary to seek brighter objects with known luminosity, and then determine the distances from their relative luminosities. Easier said than done. But in the early 1990s, it seemed that the only thing bright enough at far distances would be a supernova, so a new distance-measuring technique would have to involve supernovae. The details get really complicated from here, so let's just cut to the chase. The kind of supernovae needed was one that involved a star with known mass, which would allow its absolute brightness to be determined. Measuring its relative brightness would then yield its distance. Such an object does exist; it is called a supernova of Type Ia. So, here was the task at hand: Find supernovae of any type and sort them to find those of Type Ia. Again, easier said than done. Nevertheless, two independent teams set out to attempt these difficult, if not impossible, measurements.

Team #1: the Supernova Cosmology Project, headed by Saul Perlmutter (1959–), an astrophysicist at Lawrence Berkeley National Laboratory and professor of physics at the University of California, Berkeley. For Perlmutter's 1986 PhD project, he used an automated telescope search technique, which he extended to the supernova search. In total, the team consisted of thirty-three members.

Team #2: The High-Z Supernova Team of twenty-six members led by Brian Schmidt (1967–), then a postdoctoral student at Harvard University, which operated on the basis of three principles: be fast, be fair, and no big guns (i.e., seniority doesn't rule).

Both teams competed for precious viewing time on eight different telescopes, including the Hubble Space Telescope. They amassed data independently on many different Type Ia supernovae and determined their distances from Earth. Using the Doppler shift technique pioneered by V. M. Slipher, they determined the redshifts of the galaxies in which the supernovae were located and compared them with the Hubble relationship (see chapter 11). Both teams' results, announced within weeks of each other in 1998, showed similar results: distant supernovae were substantially dimmer than the Hubble relationship predicted. Since light from these events has taken four to eight billion years to reach us, what we're seeing is that the universe is expanding more rapidly now than it was in the past. In other words, the rate of expansion of the universe is accelerating.

IN THE DARK ABOUT DARK ENERGY

What kind of "stuff" could cause this cosmological speedup? We don't really know. In 1999, University of Chicago astrophysicist Michael Turner gave it a dark name: dark energy. This "stuff" has never been seen, so it's dark. And, since it acts in opposition to gravity, which would cause deceleration, it can't have mass in the usual sense. Using Einstein's famous $E = mc^2$, mass and energy are interconvertible, so why not call it energy. Dark energy. Catchy, eh? But this is not the only problem confronting universe theorists.

AN EARLIER DIFFICULTY

Remarkably, a huge discrepancy regarding the masses of galaxies was discovered in the 1930s, shortly after Hubble's law and Einstein's cosmological constant retraction, but this discrepancy was ignored for almost forty years. Even more remarkably, the astronomer who first noticed the problem had graduated from the ETH in Switzerland, as had Einstein, and had spent his professional career at Caltech, Mount Wilson, and Mount Palomar, as had Hubble.

Fritz Zwicky (1898–1974). Courtesy of the Archives, California Institute of Technology.

His name was Fritz Zwicky (1898–1974). Born in Bulgaria, Zwicky went to Switzerland to live with his grandparents at age six He remained a Swiss citizen all his life. Too young for World War I, Zwicky studied theoretical physics at the ETH, where he applied quantum mechanics to crystals for his PhD thesis in 1922. In 1925, Zwicky came to the United States on a Rockefeller fellowship, choosing Caltech because Pasadena's foothills bore some small resemblance to his beloved Alps. Although his sponsor, Caltech professor Robert A. Millikan, expected Zwicky to focus on quantum mechanics, Zwicky became attracted to astronomy. He began collaborating with another German-speaking astronomer, Walter Baade (1893–1960). Early in his career, Zwicky studied the cluster of galaxies known as the Coma Berenices cluster, listed by Charles Messier as M100.

Using Slipher's Doppler techniques that were later carried out at Mount Wilson by Milton Humason, Zwicky found the velocities of eight of the galaxies in the Coma cluster and estimated the mass needed to keep these galaxies gravitationally bound to the cluster. Next, he compared that mass to the cluster's mass based on the amount of light it gave off. It turned out that a lot more mass was needed to keep the cluster from flying apart. Zwicky called this invisible mass *dunkle materie*: dark matter. His calculations implied that there had to be much more dark matter than ordinary matter in the Coma cluster. Remarkably enough, this alarming result was largely ignored by other astrophysicists for almost forty years. Perhaps this was because it was published in German, "Die Rotverschiebung von Extragalaktischen Nebeln," in a sparsely read journal named *Helvetica Physica Acta*.[3]

During a long and fruitful career, Zwicky held a wide assortment of ideas of checkered quality, all pursued with relentless conviction. Some thought him brilliant, others regarded him as belligerent. Almost everyone who met Fritz Zwicky had an opinion about him. Perhaps the way he often greeted visitors to Caltech, "Who the devil are you?" should be applied to dark matter. Whatever the reason, Zwicky's dark matter didn't make a big impact on astronomy—at least not yet.

REALLY FINDING DARK MATTER

The next major contribution was made in 1970 by Vera Rubin (1928–) and W. Kent Ford (1931–). Rubin overcame many obstacles in her path,

Vera Rubin (1928–). Courtesy of the Astronomical Society of the Pacific.

but she eventually became the first woman astronomer to have access to the two-hundred-inch telescope at Mount Palomar. After working on several controversial topics, she began studying the rotation of M31 (the Andromeda Galaxy), and then more than sixty other spiral galaxies. It turned out that these galaxies were all rotating faster than their visible mass could support, again implying the existence of unseen mass. As more experimental evidence came in, the problem became too big to ignore. It seems that dark matter does exist, and there is almost ten times as much of it as ordinary bright (visible) matter.

PUTTING IT ALL TOGETHER

Thanks to a set of experiments conducted from a completely different perspective, we have an estimate of how much of this unknown dark energy and dark matter there is, even though we don't know the origin of either one. Several different experiments were designed to probe the overall geometric properties of space to determine the universe's overall makeup by looking for variations in the background microwave radiation that fills the entire universe. During the first four hundred thousand years after the big bang, the early universe was still so hot that it was opaque to electromagnetic radiation. Then it cooled sufficiently, and the radiation was emitted. During these first four hundred thousand years, the radiation could travel only a limited distance, so any fluctuations in the radiation would be limited in size. However, in traveling since then, the fluctuations would be distorted by the overall curvature of space. Measuring the size of the minute temperature fluctuations within this radiation allowed the determination of the overall curvature of space. Precise measurements have enabled the determination of the distribution of mass/energy in the entire universe. Here's the result:

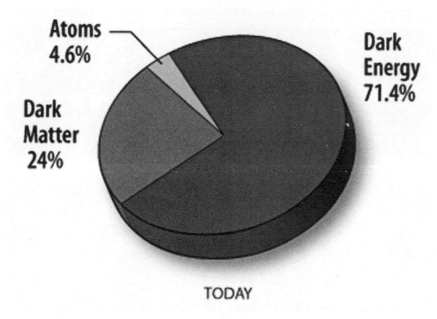

TODAY

Universe composition. From Wikimedia Commons, user CuriousMind01.

So, there it is. Most of the universe is dark energy, the next largest constituent is dark matter, and last is everything we can see: ourselves, planets, stars, galaxies, clusters of galaxies, and so on. All the visible part amounts to only 4 percent of the total universe. You may suspect the universe has gone over to the Dark Side, but we can hope that future Nobel prizewinners will sort it all out.

15.3 PARTICLE ACCELERATORS: FROM ERNEST O. LAWRENCE TO THE HIGGS BOSON AT CERN

Since the 1890s when Ernest Rutherford bombarded gold atoms with alpha particles, physics has had this "thing" about collisions: Crash things together and examine the results. Does this sound like the actions of some little kid you might know? The more energetic the collision, the more interesting products that show up in the aftermath. One such "kid," a great contributor to the development of particle accelerators, was Ernest O. Lawrence.

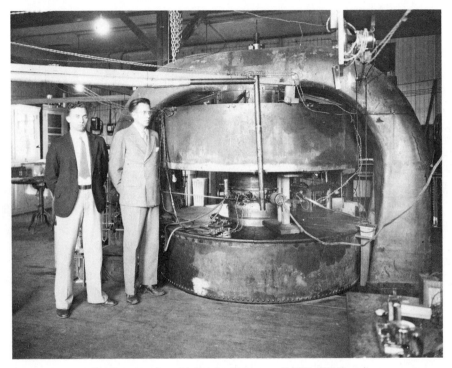

Graduate students M. Stanley Livingston (1905–1986) and Ernest O. Lawrence (1901–1958). From Wikimedia Commons, user Pieter Kuiper.

In 1928, the University of California at Berkeley lured twenty-seven-year-old Ernest O. Lawrence from Yale University, striving to build a physics department comparable to its already strong chemistry group. The following year, Lawrence, grandson of Norwegian immigrants, happened to browse a German electrical engineering journal. He saw sketches of a device proposed by Norwegian engineer Rolf Widerøe to accelerate charges to high energies by running them through an accelerating electric field, turning the particles 180° in a magnetic field, then switching the electric field direction so the particles accelerated even more. Because of the considerable engineering difficulties involved, Lawrence hesitated initially. However, he hated to lose out on the race to high energies, so, early in 1930, he assigned the task of construction of the apparatus to a graduate student, M. Stanley Livingston. By January 1931, Lawrence and Livingston had created a small working apparatus called a cyclotron. It accelerated hydrogen ions to energies of eighty thousand electron volts. In 1939, Lawrence won a Nobel Prize in Physics for the invention of the cyclotron. By 1940, twenty-two cyclotrons were either completed or under construction in the United States, and eleven more built overseas.

THE WAR INTERFERES, AGAIN

World War II halted cyclotron development for a time. When it resumed, new wrinkles were added, and the energy increased substantially. The synchrotron was developed. It featured a magnetic field that was adjusted so that as particles accelerated, their paths continued to maintain a constant radius. This allowed a smaller volume where a vacuum needed to be maintained, thus simplifying maintenance of a usable beam. Next, the synchrotron was adjusted so that particles could continue to circulate, with energy added to make up for radiation losses. This was called the storage ring. Finally, two storage rings were built adjacent to each other, and the circulating particles deflected to crash into each other. This was called the Intersecting Storage Ring configuration. It produced a great deal of basic information about fundamental particles. In the United States, the largest accelerator is the Fermi National Accelerator Lab (Fermilab), just outside Chicago. Built in 1968, this facility features a tunnel 3.9 miles in circumference. It has an accelerator called a Tevatron that can reach an energy level of 0.980 trillion electron volts (TeV) for each of its particle beams: clockwise-circulating protons and anticlockwise-circulating antiprotons. A proton-antiproton collision produces an energy of 1.96 TeV at the interaction points.

SUPERCONDUCTING SUPERCOLLIDER

Starting in 1983, a new accelerator facility was proposed. The planned energy was 20 TeV per proton, making it the largest accelerator in the world. Designed to have an underground tunnel ring of seventeen miles in circumference, the Superconducting Supercollider's (SSC) price tag was estimated in 1987 to be $4.4 billion. President Ronald Reagan approved the project in 1987, encouraging physicists to "throw deep." The project began with some controversy because of its huge size. By 1993, although only $2 billion had been spent, mostly on digging 14.6 miles of tunnel, ringing Waxahachie, Texas, the project was severely criticized by the Department of Energy's inspector general for poor management and cost overruns. Balking at the new cost estimate of $12.2 billion to complete the project, the US Congress canceled the SSC in October 1993. President Bill Clinton signed the cancelation bill, but said he regretted the "serious loss for science."[4]

ENTER CERN

Even prior to the SSC debacle, there was a complex devoted to fundamental nuclear research in Europe. It was called the European Organization for Nuclear Research, CERN. Located on the border between France and Switzerland, CERN maintained several accelerators on its site. The largest accelerator had a tunnel, built in 1988, with a circumference of 16.8 miles. This tunnel now houses the Large Hadron Collider (LHC), in which protons collided with protons at an energy of 7 TeV until March 2015, when the beam energy was updated to 15 (TeV).

HIGGS BOSON

On July 4, 2012, CERN announced the discovery of a new particle that might be the Higgs boson. Less than one year later, they confirmed this discovery, scoring a huge coup for European physics. The significance of the Higgs boson is that it confirmed the existence of a particle predicted by the Standard Model of particle physics. Although more information is needed, the existence of the Higgs boson supports the idea of the Higgs field, a fundamental part of the universe thought to be responsible for particles having mass. The upgraded beam energy may open new vistas for particle theorists. We'll see more about Higgs, the person, in mini-chapter 15.6, but the term *boson*, named after another person, is in our sights next.

15.4 BOSE, BOSONS, AND THE HIGGS BOSON

The whole edifice of modern physics is built up on the fundamental hypothesis of the atomic or molecular constitution of matter.
—C. V. Raman, Indian physicist and Nobel Laureate[5]

According to the Standard Model of particle physics, all particles are either fermions or bosons. Fermions all have half-integer spins and obey Fermi-Dirac statistics, while bosons have whole integer spins and follow Bose-Einstein statistics (to learn more about these statistics, see the To Dig Deeper section in the back matter).

Satyendra Nath Bose (1894–1974) was born in 1894 in Calcutta (now Kolkata), India. He was the firstborn, followed by six sisters. Schooling started at age five for Bose. He eventually earned his MSc at the University of Calcutta in 1915 with the highest examination scores ever recorded. Bose was fluent in Bengali, English, French, German, and Sanskrit.

Along with his colleague Meghnad Saha, Bose lectured at the University of Calcutta and translated Einstein's special and

Satyendra Nath Bose (1894–1974).
Falguni Sarkar, courtesy AIP Emilio Segre Visual Archives.

general relativity papers from German into English. In preparing a lecture for students at the University of Dhaka in 1924, Bose made a variation on the classical analysis. He treated the quanta theorized by Max Planck as indistinguishable from one another. This produced a prediction that matched the experimental evidence, whereas the classical analysis did not. Bose wrote a short article titled "Planck's Law and the Hypothesis of Light Quanta" and sent it to Albert Einstein, along with an accompanying note.

> Respected Sir, I have ventured to send you the accompanying article for your perusal and opinion. I am anxious to know what you think of it. . . . If you think the paper worth publication I shall be grateful if you arrange for its publication in *Zeitschrift für Physik* (*Journal of Physics*). Though a complete stranger to you, I do not feel any hesitation in making such a request. Because we are all your pupils though profiting only by your teachings through your writings. I do not know whether you still remember that somebody from Calcutta asked your permission to translate your papers on Relativity in English. You acceded to the request. The book has since been published. I was the one who translated your paper on Generalised Relativity[6]

Einstein found Bose's analysis quite insightful, translated it into German himself, and submitted it for publication, along with a paper of his own. The papers were well received. Bose was granted a leave from Dhaka to study in Europe, where he worked for two years with Einstein, Madame Curie, and Louis de Broglie, mostly on X-ray crystallography. Einstein adapted Bose's ideas and expanded them to become Bose-Einstein statistics.

Bose returned to Dhaka as the head of the Physics Department and later dean of the Faculty of Science until 1945. He then returned to Calcutta and taught there until he retired in 1956.

Some people thought Higgs got way more credit than Bose for the Higgs boson. But, after all, the Standard Model lists a total of thirteen bosons, but there is only one Higgs boson.

15.5 ENRICO FERMI, FRANK DRAKE, AND JILL TARTER SEARCH FOR EXTRATERRESTRIAL INTELLIGENCE

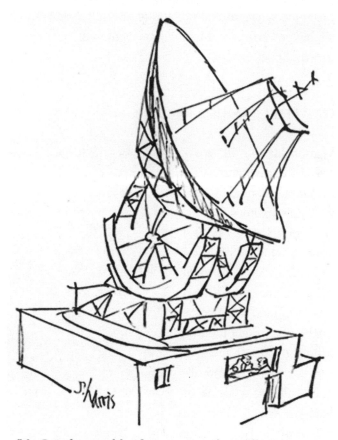

"As I understand it, they want an immediate answer. Only trouble is, the message was sent out 3 million years ago."

Used with permission from Sidney Harris.

FERMI QUESTIONS

Lunchtime with Enrico Fermi (from chapter 12) included a lot more than lunch. Conversations often featured freewheeling questions posed by

Fermi to encourage discussion (mostly by him). Thinking about all the time available for galaxy colonization by extraterrestrials, Fermi remarked, "Don't you ever wonder where everybody is?" Subsequently, this question was rephrased to "Where are they?" and is referred to as Fermi's paradox. This simple but powerful question must be dealt with by anyone interested in extraterrestrial matters.

DEALING WITH FERMI'S PARADOX

Just how probable is the exis-tence of whole extraterrestrial (ET) civilizations? National Radio Astronomy Observatory radio astronomer Frank Drake (1930–) prepared for a November 1961 informal meeting the National Academy of Sciences at Green Bank, West Virginia, to discuss extraterrestrial life. To get the dis-cussion rolling, Drake formulated an equation to focus on the prob-ability of ET life on a series of factors, each of which could be estimated separately. The equation, which Drake called the Green Bank equation, became a classic. It has been since renamed the Drake equation.

Frank Drake (1930–). Courtesy of Raphael Perrino, from Wikimedia Commons.

$$\text{Number of ET civilizations} = (\text{stars/year})(f_{planets})(f_{lifezone})(f_{life})(f_{intelligence})(f_{communicative})(\text{Lifetime})$$

Next, we'll analyze this equation term by term.

To use the Drake equation to arrive at an estimate of the number of communicative civilizations in the Milky Way Galaxy, seven factors must be estimated, where all the fs stand for fractions between 0 and 1. This is very much in the spirit of Fermi questions, so let's analyze it one factor at a time.

1. What is our galaxy's formation rate for stars suitable for generating planets suitable for life?

    Large stars have too short a lifetime, and small stars are too cool, so only middling stars need be considered here.

2. What fraction of these appropriately sized stars actually have planets?

    Given our current understanding of planet formation, it would seem most of these stars would have planets orbiting them.

3. What fraction of planets would orbit their star within a zone where life could form?

    In Earth's case, the presence of liquid water is crucial. Venus is too hot for liquid water, Mars too cold, so our solar system has just one planet in the life zone: Earth. Further, the role of the moon might be quite significant. The ebb and flow of tides could have influenced the beginnings of life here by causing pools of water to alternately flood and evaporate, possibly concentrating the "primordial soup" at critical times.

    Another unknown in life's development is the role of the massive outer planets, especially Jupiter, in deflecting possible asteroid or comet impactors out of the inner solar system. This action protected Earth from disturbing influences that might have stunted or even stopped the development of life.

4. On what fraction of appropriately placed planets does life actually arise?

    Estimating this factor usually divides optimists from pessimists. Some think that given enough carbon, liquid water, the right temperature, and enough time, life is inevitable. Others cite the myriad complexities of even a single-celled organism and say life is extremely rare, possibly even unique. Scientists differ widely on their estimates of this factor. Some doubt the usefulness of the whole approach because of this wide divergence.

5. What fraction of life-forms actually develop intelligence?

    On Earth, many species have shown evidence of intelligent behavior, humans (sometimes) included. Because intelligence seems to be such a good survival talent, it appears likely that many life-forms would develop it, given enough time.

6. What fraction of intelligent life-forms develop technologies that release detectable signals?

While both humans and dolphins are intelligent life-forms on Earth, only human technologies have generated detectable signals, so numbers like 5 percent to 50 percent are typically used for this estimate.

7. For how many years does an intelligent civilization release detectable signals into space?

This estimate provides another vehicle for expressing optimism or pessimism. An optimist might foresee a million-year civilization, whereas a pessimist might look at our own Earth's case and proclaim the end is near. Don't forget this equation was originally set up for radio astronomy purposes. A civilization could outgrow radio emissions by developing more efficient alternatives or let their radios fall into disuse as they moved on to more interesting pastimes. In our case, we have been releasing radio emissions for a little over one hundred years, so the earliest transmissions have penetrated space to a distance of one hundred light-years.

PUTTING IT ALL TOGETHER

Multiplying all these factors yields an estimate of the total number of communicative civilizations in the Milky Way Galaxy. The numbers range from billions (optimists) to only one: humans. Drake's original estimate was ten thousand. Modern versions often converge around a number of communicative civilizations approximately equal to the number of years a civilization releases detectable signals.

Although some have suggested the Drake equation is a way of encapsulating our ignorance into a small space, it is instructive to think about each of the factors in a separated format. Further, it allows another estimate to be made: the average distance between communicative civilizations. For moderate values of the seven factors above, the average distance between communicative civilizations in the Milky Way Galaxy would be hundreds to thousands of light-years. If it takes light several hundred years to travel from one civilization to another, communication would take longer than screechy old modems getting onto the Internet, if you can imagine that. Inter-civilization travel would be even less likely, at least in terms of the lifetime of a single human being. Nevertheless, for a million-year-old, technologically adept, expansionist civilization intent on colonizing the galaxy, a few thousand years of travel to a new world is not unreasonable.

Considering that the solar system has only been around for the last third of the Milky Way's existence, many other stars have quite a head start. Might they have developed the necessary technology and set out to colonize the galaxy? Knowing the galaxy's size and making reasonable assumptions about the speed of their spaceships, it would seem that such a project could be accomplished within a few million years. This is large in terms of individual human lifetimes, but small in terms of the galaxy's age. In other words, technologically advanced civilizations might very well colonize the galaxy, á la *Star Trek*, *Star Wars*, or other science fiction works.

EXPERIMENTAL ATTEMPTS

Regarding the estimates necessary to complete the Drake equation, physicist Philip Morrison said that the question we should be asking is, "Should we do something to find out?" To find the answer to the question, we must do something empirical.

Science's first experimental effort in this direction was undertaken by none other than—Frank Drake. For six hours a day, from April to July 1960, the National Radio Astronomy Observatory's eighty-five-foot dish antenna was set to 1420 MHz and pointed at two stars of about the same age as our sun. Signals from the stars Tau Ceti and Epsilon Eridani produced nothing other than a lot of static and one false alarm—a (formerly) secret military project's signals. Project Ozma, named after the queen of Oz, a character in L. Frank Baum's imaginary land "populated by strange and exotic beings,"[7] produced no positive results, but the search for ET intelligence was officially underway.

LARGE-SCALE LISTENING PROJECTS

Subsequently, several other proj-ects have been undertaken to listen for extraterrestrial transmissions. One group even sent information in case "they" were listening. The largest project, called the SETI (Search for Extraterrestrial Intelli-gence) Institute, is a not-for profit organization started in 1984. One of SETI Institute's first employees was Jill Tarter (1944–).

Tarter earned a BS in engi-neering physics from Cornell Uni-versity and a PhD in astronomy from the University of California

Jill Tarter (1944–). Courtesy of Steve Jurvetson, from Wikimedia Commons.

at Berkeley. Professionally, she has worked on a number of projects, most relating to the search for extraterrestrial life. Astronomer and cosmolo-gist Carl Sagan (more about him shortly) was associated with many of the same projects, and he began writing a screenplay, called *Contact*, partially based on Tarter's work at SETI. While the film project stalled in develop-ment, Sagan rewrote it as a novel in 1985. Eventually, the movie project was resurrected, and the movie *Contact* portrayed many aspects of SETI quite accurately, with Jodie Foster cast in a role that seems quite similar to Jill Tarter. Of course, Hollywood's search is more successful than reality.

Other search projects include optical scanners using lasers and the SERENDIP project—an acronym for Search for Extraterrestrial Radio Emissions from Nearby Developed Intelligent Populations—supported by science fiction writer Arthur C. Clarke. The project analyzes deepspacc radio tclcscopc data that it obtains while other astronomers use the telescope.

An interesting sidelight to SETI is that anyone with a computer and an Internet connection can participate in the project. You can download software that will receive data from SETI and process it when your computer is in screensaver mode. The program will display the signals, analyze them, and send the information back to SETI. See the To Dig Deeper section in the back matter for more details.

15.6 STEPHEN HAWKING AND BLACK HOLES

> Life would be tragic if it weren't funny.
>
> —Stephen Hawking[8]

Stephen Hawking's 1965 PhD thesis was titled "Singularities and Geometry of Space-Time." How appropriate for someone whose life bears so many earmarks of a singular occurrence.

Hawking was born in 1942 to a family regarded as eccentric. They got around in a reconditioned English taxi, and many meals were quiet as they were all reading books. As a youth, Hawking was recognized as bright but was not especially successful academically.

In 1959, he began college at University College, Oxford, where he found the work "ridiculously easy."[9] Eventually, he entered into the swing of college life and became interested in classical music and science fiction. He also joined the college Boat Club and served as coxswain of a rowing team that he steered along some risky

Stephen Hawking (1942–). From Wikimedia Commons, user Qz10.

routes. In his senior year, Hawking was diagnosed with ALS, or amyotrophic lateral sclerosis, which is also known as Lou Gehrig's disease and Motor Neurone Disease (MND) in the United Kingdom. For unknown reasons, brain neurons die, mostly affecting voluntary muscle function, leading to progressive muscle weakness and atrophy. Although some ALS cases are hereditary, no one else in Hawking's family has developed it. Cognitive brain centers, as well as control of elimination and eye movement, are usually spared until the end stage. Ninety percent of ALS patients die within three to four years, but typical onset occurs at age sixty. In Hawking's case, doctors estimated he would have about two to three years to live.

Remarkably, Hawking met his sister's friend Jane Wilde, and they fell in love. At their engagement, Hawking said that he now had "something to live for."[10] Within two years, Hawking finished his PhD and got married. Their first child was born two years later, with a second in three more years. Hawking taught, his physical condition deteriorated, but his fierce independence and stubbornness never wavered, even though he was finally convinced to use a wheelchair, which he drove wildly. His wife said, "Some people would call it determination, some obstinacy. I've called it both at one time or another."[11] In 1973, he wrote his first book, *The Large Scale Structure of Space-Time*, corresponding to his major research interest.

Hawking and his family spent 1974 at Caltech, during which time Jane persuaded him to accept a graduate student assistant to help with his care. Hawking's Caltech connection was very agreeable. His visits there continued for many years. In the 1970s and 1980s, he continued his study of black holes. The public had a hard time understanding black holes, partly due to their name. Perhaps they should have been called massively dense objects from which nothing, not even light, can escape. Or, in shorter version, Dense Object, No Escape, or in acronym form DONE.

When satellite observations produced evidence of black holes, namely the charged particles being squeezed together as they are about to fall into a black hole and the X-rays they emit before taking the final plunge, Hawking's work became even more interesting to the general public, and his popularity soared. During this time, his wife said her role was, in addition to the children—now three in number—and her care of Stephen, and household duties, "simply to tell him he's not God."[12] In 1988, his first attempt at a popular book, *A Brief History of Time*, was a worldwide sensation, eventually selling nine million copies.

In 1990, Hawking left his wife and moved in with one of his caregivers, Elaine Mason. Divorce and remarriage followed five years later. Although Hawking's physical condition continued to deteriorate, technological aids, such as a speech synthesizer, allowed him to maintain a vigorous professional life.

Hawking loved to make public bets with colleagues. Often, the bet would be one hundred dollars, but one time it was a year's subscription to *Penthouse* magazine. One series of bets went awry. Beginning in 1996, Hawking bet several people that the Higgs boson would not be found. After Hawking collected on several of the bets, quite publicly, normally mild-mannered Peter Higgs became annoyed, and said it was "difficult to engage him (Hawking) in discussion, so he has got away with pronounce-ments in a way that other people would not. His celebrity status gives him instant credibility that others do not have."[13]

Hawking divorced Mason in 2006. For several years prior, there had been incidents where he had been treated for injuries, but he declined to explain their source.[14]

Hollywood got into the act in 2014 with the film *The Theory of Everything*, which garnered a Best Actor Academy Award for Eddie Redmayne, who played the role of Hawking.

The film, based on Jane Hawking's memoir, *Travelling to Infinity*, was accepted by Hawking as "broadly true."[15] However,

Eddie Redmayne. Courtesy of Steve Jurvetson, from Wikimedia Commons.

his sisters and second ex-wife boycotted the premiere, citing factual inaccuracies.

In July 2015, the Breakthrough Funding Program was instituted. It is a ten-year, hundred-million-dollar initiative to inject funds into the SETI program. Funded by Russian tycoon Yuri Milner, it will feature the most extensive alien communication program to date. Milner was a PhD candidate in particle physics, but he changed fields to earn an MBA from the Wharton School of Business after being "disappointed in myself as a physicist."[16] During the project launch, at London's Royal Society, Hawking said, "In an infinite Universe, there must be other life. There is no bigger question. It is time to commit to finding the answer."[17] Hawking and Milner were joined by Frank Drake (see the previous mini-chapter) and Ann Druyan (next mini-chapter) at the announcement in London. Milner regards SETI as a "low probability high impact project," and he's the one who supplied the money.[18]

15.7 COSMOS–CARL SAGAN + ANN DRUYAN + NEIL DEGRASSE TYSON + SETH MACFARLANE

> For me, it is far better to grasp the Universe as it really is than to persist in delusion, however satisfying and reassuring.
> —Carl Sagan[19]

> My knowledge of science came from being with Carl, not from formal academic training. Carl gave me a thrilling tutorial in science and math that lasted the 20 years we were together.
> —Ann Druyan[20]

> I said that if an alien came to visit, I'd be embarrassed to tell them that we fight wars to pull fossil fuels out of the ground to run our transportation. They'd be like, "What?"
> —Neil deGrasse Tyson[21]

> The continuance of our journey outward into space should always occupy some part of our collective attention, regardless of whatever Snooki did last week.
> —Seth MacFarlane[22]

"Billions and billions . . ." That's what many people think when they hear the name *Carl Sagan*. But that phrase actually annoyed Sagan himself. He wrote to Johnny Carson, protesting that he didn't actually say that in his many appearances on the *Tonight Show*. The response: "Even if you didn't say 'billions and billions' you should have—Johnny."[23] Actually, the phrase played right into Sagan's hands. His quest to make science, especially astronomy, popular and understandable required people scaling up their thinking to billions and billions. But his life didn't start out quite that way.

Carl Sagan (1934–1996). From Wikimedia Commons, user Erlendaakre.

Born in the Bensonhurst neighborhood of Brooklyn in 1934, Sagan said, "My parents were not scientists. They knew almost nothing about science. But in introducing me simultaneously to skepticism and to wonder, they taught me the two uneasily cohabiting modes of thought that are central to the scientific method."[24] From the University of Chicago, Sagan achieved a BS and MS in physics and a PhD in astronomy and astrophysics with a 1960 dissertation on "Physical Studies of Planets and Satellites." During a two-year fellowship at the University of California, Berkeley, Sagan predicted that Venus was dominated by a carbon dioxide greenhouse effect, and had a very high surface temperature, contradicting earlier ideas of a cool surface. This prediction was confirmed by later spacecraft measurements.

GO EAST, YOUNG MAN

Sagan then moved to Massachusetts and worked at the Smithsonian Astrophysical Observatory at Cambridge. He also taught and did research at

Harvard but was denied tenure in 1968. In 1969, he wrote an essay under the pseudonym Mr. X about the use of cannabis for the 1971 book *Marihuana Reconsidered* by his good friend and fellow Harvard academic (in psychology) Lester Grinspoon. He became full professor at Cornell in 1971 and directed the Laboratory for Planetary Studies there. In 1973 Sagan wrote, with Jerome Agel, *The Cosmic Connection: An Extraterrestrial Perspective*. This book, and all the other twenty-plus popular books he wrote were actually dictated by Sagan and transcribed into type. Sagan would then edit them repeatedly, by hand, on their way to being printed.

In the sixties and seventies, Sagan consulted on many NASA spacecraft missions and often used his communication skills to explain things to reporters. He became a favorite of Johnny Carson, appearing as almost a resident astronomer on the *Tonight Show*. The book *Cosmos* and the PBS series based on the book, *Cosmos: A Personal Voyage*, brought instant fame and recognition to Sagan.

SELF-APPRAISAL

Sagan almost always seemed like the smartest person in the room, but he had this to say about himself: "I think I'm able to explain things because understanding wasn't entirely easy for me. Some things that the most brilliant students were able to see instantly I had to work to understand."[25] He wrote many articles that appeared in the popular press and listened patiently to the unusual opinions of others, even about UFOs. But after a thorough hearing, Sagan always applied the scientific method to all claims, pointing out that "extraordinary claims require extraordinary evidence."[26] His last book even featured a chapter titled "The Fine Art of Baloney Detection." Some of the more eccentric letters he received were labeled "F/C," Sagan's shorthand for fractured ceramics—crackpots.

SAGAN COLLABORATOR

One of Sagan's cowriters on *Cosmos* was author and TV producer Ann Druyan.

Druyan was born in 1949 and grew up in Hollis, Queens, New York City. She was a dropout from NYU in the late sixties and met Sagan at a party (hosted by journalist Nora Ephron) in New York City, Sagan with his second wife, Druyan with another man. They worked on several TV projects that fizzled before Sagan asked her to become a creative director on the

Ann Druyan. Courtesy of Bob Lee, from Wikimedia Commons.

Voyager space project he was working on. She helped design the Voyager Golden Records, a collection of images, music, and sounds included on the *Voyager 1* and *2* space probes, which are now, literally, billions and billions of miles away. Druyan became Sagan's third wife in 1981.[27] They collaborated on writing *The Demon-Haunted World*, *Comet*, and *Shadows of Forgotten Ancestors*. Together, they conceived the story that would eventually become the novel and film *Contact*, for which Druyan served as producer. During production of the movie, Sagan died of pneumonia resulting from myelodysplastic syndrome (MDS) at the age of sixty-two. Filming was suspended briefly, then the movie was finished and released to critical acclaim in 1997.

At Cornell, Sagan often met with potential students to convince them to enroll at Cornell. In 1975, he interviewed a Bronx High School of Science senior who had evidenced a substantial interest in the universe and had even given astronomy lectures at the age of fifteen. This young man's name was Neil deGrasse Tyson. Here's a photo taken a bit later in his life:

Tyson spent a day touring Cornell with Sagan, who had offered to put him up overnight if his bus didn't come. Tyson recalled, "I already knew

Selfie with Bill Nye (1955–), President Barack Obama (1961–), and Tyson (1958–).
From Wikimedia Commons, user Stemoc.

I wanted to become a scientist. But that afternoon, I learned from Carl the kind of *person* I wanted to become."[28] In spite of Sagan's recruitment efforts, Tyson attended Harvard and got a BS in physics there in 1980. He also rowed, wrestled, and participated in campus dance activities. Tyson continued his studies at the University of Texas at Austin, earning an MS in astronomy. He then went to Columbia, where he was a coauthor of a paper with Brian Schmidt early in the Supernova Ia project (recall chapter 15.2). Tyson got his PhD in astrophysics from Columbia in 1991.

He joined the Hayden Planetarium staff and became director in 1996. According to him, "When I was a kid . . . there were scientists and educators on the staff at the Hayden Planetarium . . . who invested their time and energy in my enlightenment . . . and I've never forgotten that."[29] Tyson has written many books, served on many boards and panels, and appeared on many TV programs such as the *Daily Show* and *Colbert Report*, all to explain and popularize astronomy and science. He also served in many capacities on the Planetary Society, an organization founded in 1980 by Sagan.

Tyson and Druyan had talked for a while about re-creating *Cosmos* for TV, but the project finally began to get somewhere when Tyson introduced Druyan to Seth MacFarlane in 2008.

MacFarlane (1973–) is a multitalented artist who started as an animator but expanded his career to become an actor, singer, producer, and director. When MacFarlane learned of the efforts to make a new version of *Cosmos*, he was excited. He told Tyson, "I'm at a point in my career where I have some disposable income . . . and

Seth MacFarlane (1973–). Courtesy of Gage Skidmore, from Wikimedia Commons.

I'd like to spend it on something worthwhile."[30] MacFarlane set up meetings at Fox Network and helped Druyan and Tyson sell the project. *Cosmos: A Spacetime Odyssey* premiered in March 2014. It was a thirteen-episode series like the original *Cosmos*, and had Seth MacFarlane and Ann Druyan among its executive producers. Similar to the original *Cosmos: A Personal Voyage*, it featured graphic elements such as the "Ship of the Imagination" and the "Cosmic Calendar," but all the animations were substantially upgraded. The show has received several awards, including a Peabody for educational content. Some people wondered if Tyson was trying to fill Sagan's shoes. His response was, "If I try to fill his shoes I'll just fail, because I can't be him, but I can fill my own shoes."[31] Tyson wears size twelve and a half shoes, which appear pretty substantial and well filled.

15.8 GEORGE WASHINGTON CARVER

There is no short cut to achievement. Life requires thorough preparation—veneer isn't worth anything.
—George Washington Carver[32]

George Washington Carver (1864–1943).
From Wikimedia Commons, user Scewing.

Although he has been called "the Peanut Man," there's a lot more to George Washington Carver (1864–1943) than peanuts. Carver was born into slavery near Diamond, Missouri, sometime around 1864. His sister, his mother, and he were kidnapped by night raiders from Arkansas when he was only about a week old. Moses Carver, George's master, hired a tracker to find them, but only George was found. After the war, George and his older brother James were raised by Moses and Susan Carver as their own children, with "Aunt Susan" teaching him to read and write.

George had an abiding love of nature and a deep-seated curiosity. He seemed to have an instinctive knowledge about growing things. After curing many ailing plants, George became known as "the Plant Doctor." When he was about twelve, with the Carver's blessings, George set out for Neosho, Missouri, where there was a school that accepted black students. There, George boarded with Mariah and Andrew Watkins, who soon grasped the power of George's mind. "You must learn all you can and then go out into the world and give your learning back to our people that's so starving for a little learning."[33] Mariah Watkins also told him to stop calling himself "Carver's George," because he was no one's property.

Within a year, George Carver moved on from Neosho, having learned all he could. After various odd jobs and occasional academic work, George finished high school in Minneapolis, Kansas, in 1885. To avoid confusion with another George Carver, he became George W. Carver. When asked if the *W* stood for "Washington," he said why not. In addition to knowing about plants, Carver enjoyed painting pictures of them, using homemade brushes and paints.

Carver became the first black student at Iowa State Agricultural College at Ames, Iowa, in 1891. He earned both bachelor's and master's degrees in botany, and achieved national recognition for his research in plant pathology and mycology (study of fungi). Carver then began teaching and became Iowa State's first black faculty member.

TUSKEGEE TIMES

Booker T. Washington (1856–1915).
From Wikimedia Commons, user Ineuw.

Much farther east and south, another school employed almost exclusively black faculty. The Tuskegee Normal and Industrial Institute was founded in 1881, and its first faculty member was Booker T. Washington (1856–1915).

Washington was a recent graduate of Hampton Normal and Agricultural Institute. His goal was to train black teachers to build the black community's economic strength and pride by emphasizing self-help and schooling, especially in agricultural matters. As Tuskegee Institute flourished, Washington became a dominant leader in the black community. He believed they should "concentrate all their energies on industrial education, and accumulation of wealth, and the conciliation of the South."[34] The goal was to overcome white prejudice slowly by showing that blacks deserved equality on the basis of their merits.

As part of his efforts to strengthen the faculty, Washington invited George Washington Carver to be head of the Agriculture Department in 1896. In Washington's letter, he said Carver would have "the challenge of bringing people from degradation, poverty and waste"[35] to more fruitful lives. Although he was considering an offer from Alcorn Agricultural and Mechanical College in Mississippi, Carver admired Washington and decided to leave his comfort zone in the West to travel to the heart of the South.

When Carver arrived in Alabama, he saw "not much evidence of scientific farming anywhere. Everything looked hungry: the land, the cotton, the cattle and the people."[36] Carver taught that the way to renew the soil and the people would be to plant crops largely ignored in the South: sweet potatoes, black-eyed peas, and peanuts. All three were both soil-renewing and edible.

In the almost twenty years that Washington and Carver worked together at Tuskegee Institute, they had many a rocky moment. Carver's responsibilities were so diverse that he was spread thin and his paperwork lagged far behind his performance. On the other hand, Washington never seemed to provide enough funds for the equipment that Carver requested. This may sound typical to anyone familiar with academic politics, yet Carver threatened to resign several times because of disputes. Washington always managed to smooth things over, often just in time. Even bigger problems for Washington were posed by the more activist W. E. B. Du Bois (1868–1963), who challenged Washington's go-slow policy with a far more militant leadership style for the black community.

Yet, both Washington and Carver maintained their mutual respect. In 1911, Washington called Carver "one of the most thoroughly scientific men of the Negro race with whom I am acquainted."[37] When Washington died after a short illness in 1915, Carver donated half his year's salary to the Booker T. Washington Memorial Fund.

With Washington gone, Carver worked on, researching new uses of peanuts, soybeans, sweet potatoes, pecans, and other crops. His appearance before the Peanut Growers Association in 1920 and testimony before the US Congress supporting a tariff on imported peanuts led to national prominence and consultation with US presidents Theodore Roosevelt and Franklin Delano Roosevelt.

Carver was awarded an honorary doctorate from Simpson College in 1928. In 1941, *Time* magazine dubbed Carver a "Black Leonardo." That's a pretty good honor for a Peanut Man.

15.9 HEDY LAMARR AND GEORGE ANTHEIL–UNEXPECTED FREQUENCY HOPPING

All creative people want to do the unexpected.

—Hedy Lamarr[38]

Inventive people show up in all manner of shapes, sizes, and appearances. The odd couple in this mini-chapter would appear to be about as far removed from weapons electronic security designers as one could imagine, but their US Patent 2292387 dealt with weapons security. According to its description, "This invention relates broadly to secret communication systems involving the lie of carrier waves of different frequencies and is especially useful in the remote control of dirigible craft, such as torpedoes."[39] The year was 1942, and the War Office classified the patent immediately as top secret. It was granted to Hedy Kiesler Markey of Los Angeles and George Antheil of Manhattan Beach, California. Who are these two people, and how did they link up?

Hedy Lamarr (1913–2000). From Wikimedia Commons, user Rossrs~commonswiki.

The woman listed on the patent as Hedy Kiesler Markey (Gene Markey was her second husband out of six total) was born Hedwig Eva Maria Kiesler on November 9, 1913, in Vienna, Austria. Hedwig was an only child. Her father was an affluent banker, and she studied ballet and classical piano before entering Max Reinhardt's famous Berlin acting school. Hedy (the nickname for Hedwig) appeared in a few films as a teenager, most notably the Czech film *Ecstasy*. Her role involved a brief nude scene and simulated sex. While tame by twenty-first-century standards, this was very daring in 1933, so the film was banned in Germany and the United States.

Shortly after the film was released, Hedy married Fritz Mandl, who was director general of weapons manufacturer Hirtenberger Patronenfabrik. The marriage was not a happy one, and Hedy often found herself in the company of munitions experts who discussed the latest German military research. She soon realized the connection this group had to the Nazi war machine, and it frightened her. Hedy's husband, ever jealous, tried to buy up and destroy copies of *Ecstasy* but failed. After putting up with the demands of this unhappy marriage for almost four years, Hedy made an elaborate plan to escape Mandl's clutches and left carrying only a few jewels and a small suitcase.

She linked up with Louis B. Mayer, signed a movie contract, and sailed to the United States aboard the *Normandie*. By the time she landed, she had a new name, Hedy Lamarr, and a new career stretching before her. Max Reinhardt had called her "the most beautiful woman in the world,"[40] and the MGM publicity department embellished that title. In her first American film, *Algiers*, with Charles Boyer, this new actress's beauty took everyone's breath away. She made a succession of films with Clark Gable, Jimmy Stewart, Spencer Tracy, and Victor Mature, but some critics think MGM didn't utilize her talents fully, suspecting that Mayer never quite understood how to handle her independence.

Lamarr didn't fit the Hollywood mold. Shooting several films per year left her with a substantial amount of "down time." Celebrity parties didn't really interest her; she actually preferred smaller gatherings with friends. Also, she had a drafting table set up at home and enjoyed working on inventions, such as a cube that would produce an instant soft drink when mixed with water.

At a small party given by actress Janet Gaynor, Lamarr met composer George Antheil. She had read one of Antheil's *Esquire* articles about glands and wanted to discuss how glands might increase her breast size. That conversation didn't go very far, but the two imaginative and inventive minds found many more fruitful subjects.

ENTER ANTHEIL

George Antheil (1900–1959) was an American composer of avant-garde music and writer of magazine articles, newspaper columns, and even a mystery novel. One of his best-known works was *Ballet Mécanique*, originally scored for sixteen synchronized player pianos, two grand pianos, electronic bells, xylophones, bass drums, a siren, and three airplane propellers. By the late 1930s, Antheil was writing film scores for Hollywood and was delighted to meet Hedy Lamarr.

George Antheil (1900–1959). From Wikimedia Commons, user Stef joosen.

Lamarr and Antheil spent countless hours drafting and redrafting their torpedo design and other inventions, including a proximity fuse for antiaircraft shells. According to their patent application, to accomplish frequency changes for their torpedo, "we contemplate employing records of the type used for many years in player pianos, and which consist of long rolls of paper having perforations variously positioned in a plurality of longitudinal rows along the records. In a conventional Player Piano record there may be 88 rows of perforations. And in our system such a record would permit the use of 88 different carrier frequencies, from one to another of which both the transmitting and receiving station would be changed at intervals."[41]

Although the design wasn't used by the War Office, the idea caught hold later and is now considered the foundational patent for spread spectrum technologies used in cell phones, Bluetooth, and GPS devices.

In 1997, Hedy Lamarr was presented the Electronic Frontier Foundation (EFF) Pioneer Award. At the award ceremony, EFF staff counsel Mike Godwin said, "Lamarr and Antheil had hoped that the military applications of their invention would play a role in the defeat of Nazi Germany. Ironically, this tool they developed to defend democracy half a century ago promises to extend democracy in the twenty-first century."[42] If there had been an award for "Most Beautiful Frequency Hopping," she might have won that, too.

15.10 THE AMAZING RANDI

The New Age? It's just the old age stuck in a microwave oven for fifteen seconds.

—James Randi[43]

Last, but certainly not least on the list of honorable mentions, is a man who is technically not a scientist. Actually, he dropped out of high school at age seventeen. And yet he has performed an extremely valuable function for science. Perhaps an analogy will help to explain his actions.

A sculptor was once asked how he managed to create such a wonderful likeness of a heroic figure. He said it was very simple: Just chip away anything that doesn't look heroic.

A major challenge faced by science is chipping away ideas that might seem to be scientific but that actually undermine the whole structure of science by passing off illusions as if they are reality. Educators are familiar with the crust of nonscientific ideas that must be chipped off before real science can enter people's minds. Carl Sagan treated this problem in a general way in his book chapter "Baloney Detection Kit" (see chapter 15.7) but the subject of this mini-chapter concerns himself primarily with paranormal, occult, and supernatural claims. He refers to these items collectively as "woo-woo."[44] Our final honoree is The Amazing Randi.

THE AMAZING RANDI

Randall James Hamilton Zwinge was born in Toronto, Ontario, Canada, in 1928. As a youth, a bicycle accident put him into a full-body cast for thirteen months, but he learned to walk again, much to the surprise of

"HOLD IT. THEY DECIDED THEY'LL JUST BUY THE 9 INCH SCALE MODEL."

Used with permission from Sidney Harris.

his doctors. Seeing a performance by magician Harry Blackstone Jr. and reading books about magic inspired him to drop out of school at seventeen to become a conjurer in a traveling road show. Eventually calling himself "The Amazing Randi," he performed in nightclubs, posed as a psychic, and briefly wrote a column on astrology for the tabloid *Midnight* under the name of Zo-ran. His technique was quite simple: he took items from other newspaper astrology columns, shuffled them randomly, and pasted them into his column.

The Amazing Randi (1928–). Courtesy of S. Pakhrin, from Wikimedia Commons.

In the fifties, Randi worked in the United Kingdom, Europe, the Philippines, and Japan. Many of his fellow performers and some evangelists claimed to have supernatural powers, but Randi could tell what they were actually doing. One was the "go-ahead" or "billet reading" technique. Audience members write a statement on paper that is then sealed into an envelope. One of the audience members is a plant, who writes something that the performer has memorized beforehand. The plant then marks the envelope. When the performer has collected all the envelopes, he puts the marked one on the bottom of the stack. The performer selects an envelope from the top of the stack, goes through his magic motions, and announces the preplanned statement. The plant loudly proclaims its accuracy, the performer opens the envelope, and says he got it right. Of course, the envelope opened actually reveals someone else's statement, written by a real audience member, which the performer memorizes. He then selects another envelope and states the contents he memorized from the prior envelope. Randi performed these tricks and other mind-reading illusions but became discouraged by the public acceptance of them as being paranormal. Randi then shifted his focus and began to perform more escape tricks, similar to Harry Houdini. Once he spent fifty-five minutes in a block of ice. Another time, he escaped from a straitjacket while suspended over Niagara Falls.

ENTER URI GELLER

Randi entered the public eye when he issued a public challenge to Uri Geller, a former Israeli paratrooper. Geller had claimed paranormal powers, but Randi said he used standard magic tricks to bend spoons and perform other feats of telekinesis. It wasn't that Randi objected to magic tricks—that's what he did for a living. What bothered him was that Geller fooled gullible people into believing in supernatural powers. In 1976, the Committee for the Scientific Investigation of Claims of the Paranormal (CSICOP) was formed, with the founding board members psychologist Ray Hyman, *Scientific American* columnist Martin Gardner, secular humanist philosopher Paul Kurtz, and authors Isaac Asimov and Carl Sagan (see chapter 15.7) as well as Randi. Part of the funding came from the sale of a new magazine named the *Skeptical Inquirer*. In his role as paranormal investigator, Randi debunked a wide variety of frauds, but he was careful to be called an investigator rather than a debunker. "Because if I were to start out saying, 'This is not true, and I'm going to prove it's not true,' that means I've made up my mind in advance. So every project that comes to my attention, I say, 'I just don't know what I'm going to find out.' That may end up—and usually it does end up—as a complete debunking. But I don't set out to debunk it."[45]

Randi's pursuit of Geller continued with appearances in person, on TV (the *Tonight Show* with Johnny Carson), and even a book, *The Truth about Uri Geller*. Eventually, Geller responded by suing Randi and CSICOP for fifteen million dollars. Although the suit was thrown out, potential legal costs led CSICOP's leadership to demand that Randi stop investigating Geller. Randi resigned from CSICOP, and the group changed its name to Committee for Skeptical Inquiry (CSI). After Geller's suit was thrown out of court, Randi joined the board of CSI.

The James Randi Educational Foundation (JREF) was started in 1996 to "help people defend themselves from paranormal and pseudoscientific claims. The JREF offers a still-unclaimed million-dollar reward for anyone who can produce evidence of paranormal abilities under controlled conditions."[46] The One Million Dollar Paranormal Challenge has been hosted at an annual gathering of academics, comedians, magicians, and writers billed as The Amazing Meeting. So far, no claimants have been awarded the prize, but several have tried. (Full disclosure: One of us, CMW, has hosted Randi at his university and attended and presented at several Amazing Meetings.)

The Amazing Randi has had a long and fascinating career, often appearing on TV, including a *60 Minutes* exposé using a fake "spirit channeler" featuring a friend of Randi's who eventually became Randi's life partner. To find out more about Randi, visit the To Dig Deeper section in the back matter.

Although he wasn't trained as a scientist, Randi's powers of observation probably exceed many scientists', and his open-mindedness in evaluating what he investigates can serve as a model of objectivity for all science.

AFTERWORD

Used with permission from Sidney Harris.

After you've lived in a house for a while, you develop a vivid mental picture of where you live. But a curious thing happens if you sit on a neighbor's patio and look back at your house. This new perspective allows you to see your house a bit differently, and your mental image of it gets revised.

So it is with this book. As we authors viewed science from the standpoint of the humanity of its participants, our mental picture of how scientific ideas evolve has changed markedly. In this book, we have chronicled

Used with permission from Sidney Harris.

almost four hundred people's interactions over twenty-five hundred years and in dozens of countries of the world. Influenced by person-to-person interactions, these people produced big ideas of science. These interactions were often cooperative, sometimes contentious, and frequently connected, reinforcing the idea of us living on a "small world." Thanks to the refining power of the scientific method, only those ideas that continued to be supported by physical evidence survived the evaluation process and became truly big ideas.

As to the unsolved problems science faces today: dark energy, dark matter, DNA details, the Theory of Everything ... "The more things change, the more they stay the same."[1] The next group of Newtons, Einsteins, Teslas, and Hubbles have different challenges, but they, too, will be influenced by human interactions that will continue to be cooperative, contentious, and connected.

How could we expect it to be any different?

NOTES

PREFACE

1. "Craig Venter Quotes," BrainyQuote.com, accessed September 2015, http://www
.brainyquote.com/quotes/quotes/c/craigvente554125.html.

2. "Francis Collins Quotes," BrainyQuote.com, accessed September 2015, http://
www.brainyquote.com/quotes/quotes/f/franciscol555127.html.

INTRODUCTION

1. "Albert Einstein Quotes," BrainyQuote.com, accessed September 2015, http://
www.brainyquote.com/quotes/quotes/a/alberteins100017.html.

2. *Wikipedia*, s.v. "George Francis FitzGerald," last modified December 3, 2015,
https://en.wikipedia.org/wiki/George_Francis_FitzGerald.

3. Edward Neville Da Costa Andrade, *Rutherford and the Nature of the Atom* (New
York: Doubleday, 1964), p. 111.

CHAPTER 1: DEMOCRITUS AND ARISTOTLE PONDER
THE EXISTENCE OF ATOMS

1. Democritus, as quoted by Diogenes Laërtius, "Democritus Quotes," Quotes,
Your Dictionary, accessed November 2015, http://quotes.yourdictionary.com/author/
democritus/154834.

2. Democritus, as quoted by Diogenes Laërtius in Alan L. Mackay, *A Dictionary of
Scientific Quotations* (Boca Raton, FL: CRC, 1992, 1994), p. 71.

3. Aristotle, *Physics*, book 2, part 3, trans. P. Hardie and R. K. Gaye, Internet Classics
Archive, accessed September 2015, http://classics.mit.edu/aristotle/history_anim.2.ii.html.

4. Simplicius, commenting on Aristotle's *Physics* in G. S. Kirk, J. E. Raven, and M.
Schofield, eds., *The Pre-Socratic Philosophers* (Cambridge: Cambridge University Press,
1983), p. 472.

5. Aristotle, *The Works of Aristotle*, trans. Sir David Ross (Oxford, UK: Clarendon,
1952), vol. 12, Archive.org, accessed September 2015, https://archive.org/stream/works
ofaristotle12arisuoft/worksofaristotle12arisuoft_djvu.txt.

CHAPTER 2: ARISTOTLE, ARISTARCHUS, COPERNICUS, AND GALILEO SEEK TO DETERMINE EARTH'S PLACE IN THE COSMOS

1. "Carl Sagan Quotes," BrainyQuote.com, accessed September 2015, http://www.brainyquote.com/quotes/quotes/c/carlsagan133582.html.

2. Archimedes, *The Sand Reckoner* in T. L. Heath, *Aristarchus of Samos, the Ancient Copernicus: A History of Greek Astronomy to Aristarchus Together with Aristarchus' Treatise on the Sizes and Distances of the Sun and Moon, a New Greek Text with Translation and Notes* (Oxford, UK: Clarendon, 1913), p. 302.

3. *Wikipedia*, s.v. "Tycho Brahe," last modified December 13, 2015, https://en.wikipedia.org/wiki/Tycho_Brahe#Tycho.27s_nose.

4. Stephen Webb, *Measuring the Universe: The Cosmological Distance Ladder* (Berlin: Springer Praxis, 1999), p. 38.

5. *Wikipedia*, s.v. "Tycho Brahe."

6. Johannes Kepler, *Astronomia Nova*, chap. 149 (Heidelberg, Germany, 1609), pp. 113–14.

7. Anton Postl, "Correspondence between Kepler and Galileo," *Vistas in Astronomy* 21 (1977): 325–30, ScienceDirect, accessed September 2015, http://www.sciencedirect.com/science/article/pii/0083665677900204.

8. Galileo Galilei, *Dialogue concerning the Two Chief World Systems*, transl. Stillman Drake (Berkeley: University of California Press, 1953; rev., 1967).

9. *Wikipedia*, s.v. "And Yet It Moves," last modified December 28, 2015, https://en.wikipedia.org/wiki/And_yet_it_moves.

CHAPTER 3: ISAAC NEWTON, ROBERT HOOKE, AND GOTTFRIED LEIBNIZ ARGUE ABOUT MOTION AND CALCULUS

1. *Wikipedia*, s.v. "Early Life of Isaac Newton," last modified December 1, 2015, https://en.wikipedia.org/wiki/Early_life_of_Isaac_Newton.

2. Ibid.

3. *Wikiquote*, s.v. "Isaac Newton," last modified December 1, 2015, https://en.wikiquote.org/wiki/Isaac_Newton.

4. *Wikipedia*, s.v. "Robert Hooke," last modified November 16, 2015, https://en.wikipedia.org/wiki/Robert_Hooke.

5. Lisa Jardine, *The Curious Life of Robert Hooke: The Man Who Measured London* (New York: HarperCollins, 2005), pp. 5–10.

6. Tim Sharp, "Edmond Halley Biography: Facts, Discoveries, and Quotes," Space

.com, February 12, 2014, accessed September 2015, http://www.space.com/24682
-edmond-halley-biography.html.

7. *Wikipedia*, s.v. "Robert Hooke."

8. "Memoirs of the Life, Writings, and Discoveries of Sir Isaac Newton," Archive
.org, accessed September 2015, http://archive.org/stream/aat0604.0001.001.umich.edu/
aat0604.0001.001.umich.edu_djvu.txt.

9. Jardine, *The Curious Life of Robert Hooke*, pp. 15–18.

10. "Flamsteed, John," in *Complete Dictionary of Scientific Biography*, Charles Scribner's Sons, Encyclopedia.com, accessed November 2015, http://www.encyclopedia.com/
topic/John_Flamsteed.aspx.

11. *Wikipedia*, s.v. "Gottfried Wilhelm Leibniz," last modified December 20, 2015,
https://en.wikipedia.org/wiki/Gottfried_Wilhelm_Leibniz.

12. *Wikipedia*, s.v. "Leibniz–Newton Calculus Controversy," last modified December
21, 2015, https://en.wikipedia.org/wiki/Leibniz–Newton_calculus_controversy.

13. *Wikiquote*, s.v. "Isaac Newton."

CHAPTER 4: THE BATTLING BERNOULLIS AND
BERNOULLI'S PRINCIPLE

1. "Daniel Bernoulli Quotes," Wise-Quote.com, accessed November 2015, http://
wise-quote.com/Daniel-Bernoulli-8317.

2. *Wikipedia*, s.v. "Brachistochrone Curve," last modified July 25, 2015, https://
en.wikipedia.org/wiki/Brachistochrone_curve.

3. Richard S. Westfall, *Never at Rest: A Biography of Isaac Newton* (Cambridge:
Cambridge University Press, 1983), p. 582.

4. Ibid., p. 583.

5. Ibid.

6. Ibid.

7. Carl B. Boyer, *A History of Mathematics* (New York: Wiley, 1968).

CHAPTER 5: ANTOINE LAVOISIER AND
BENJAMIN THOMPSON (COUNT RUMFORD)
HAVE RIVAL THEORIES OF HEAT

1. "Harry S. Truman Quotes," BrainyQuote.com, accessed September 2015, http://
www.brainyquote.com/quotes/quotes/h/harrystrum162028.html.

2. *Science Newsletter* 14 (1928): 52.

3. *Wikiquote*, s.v. "Joseph Louis Lagrange," last modified October 29, 2015, https://
en.wikiquote.org/wiki/Joseph_Louis_Lagrange.

4. Robert T. Grimm, *Notable American Philanthropists: Biographies of Giving and Volunteering* (Westport, CT: Greenwood, 2002), p. 320.

5. Christa Jungnickel and Russell McCommach, *Cavendish* (Philadelphia: American Philosophical Society, 1996), p. 351.

6. Ioan James, *Remarkable Physicists from Galileo to Yukawa* (Cambridge: Cambridge University Press, 2004), p. 85.

7. William Gurstelle, *The Practical Pyromaniac* (Chicago: Chicago Review Press, 2011), p. 55.

CHAPTER 6: MENDELEEV, MEYER, MOSELEY AND THE BIRTH OF THE PERIODIC TABLE

1. *Wikiquote*, s.v. "Harlan Ellison," last modified December 20, 2015, https://en.wikiquote.org/wiki/Harlan_Ellison.

2. *Wikipedia*, s.v. "*The Sceptical Chymist*," last modified December 25, 2015, https://en.wikipedia.org/wiki/The_Sceptical_Chymist.

3. "Antoine Lavoisier," Chemistry Explained, accessed January 6, 2016, http://www.chemistryexplained.com/Kr-Ma/Lavoisier-Antoine.html.

4. Dmitry Ivanovich Mendeleev, *Principles of Chemistry* (Minneapolis: Filiquarian Legacy, 2012).

5. "Dmitrii Ivanovich Mendeleev," in *Complete Dictionary of Scientific Biography*, Charles Scribner's Sons, Encyclopedia.com, accessed January 6, 2016, http://www.encyclopedia.com/topic/Dmitrii_Ivanovich_Mendeleev.aspx.

6. "Henry Moseley," Famous Scientists, December 29, 2014, accessed January 6, 2016, http://www.famousscientists.org/henry-moseley/.

7. Ibid.

CHAPTER 7: WESTINGHOUSE AND TESLA VERSUS EDISON— AC/DC TITANS CLASH

1. "Nikola Tesla Quits Working for Edison," World History Project, accessed September 2015, https://worldhistoryproject.org/1885/nikola-tesla-quits-working-for-edison.

2. *Wikiquote*, s.v. "Thomas Edison," last modified December 10, 2015, https://en.wikiquote.org/wiki/Thomas_Edison.

3. *Wikiquote*, s.v. "Nikola Tesla," last modified December 5, 2015, https://en.wikiquote.org/wiki/Nikola_Tesla.

4. *Wikipedia*, s.v. "War of Currents," last modified December 24, 2015, https://en.wikipedia.org/wiki/War_of_Currents.

5. Mark Essig, *Edison and the Electric Chair: A Story of Light and Death* (New York: Bloomsbury, 2009), p. 136.

6. *Wikipedia*, s.v. "Edison Ore-Milling Company," last modified October 18, 2015, https://en.wikipedia.org/wiki/Edison_Ore-Milling_Company.

CHAPTER 8: ALFRED WEGENER STANDS HIS GROUND ABOUT CONTINENTAL DRIFT

1. "Mary Roach Quotes," BrainyQuote.com, accessed September 2015, http://www.brainyquote.com/quotes/quotes/m/maryroach694698.html.

2. *Wikipedia*, s.v. "Continental Drift," last modified December 28, 2015, https://en.wikipedia.org/wiki/Continental_drift.

3. Ibid.

4. "Excerpts and Readings on Alfred Wegener (1880–1930)," Pangaea, accessed January 6, 2016, http://pangaea.org/wegener.htm.

CHAPTER 9: PART 1: ALBERT EINSTEIN, MARCEL GROSSMANN, MILEVA MARIĆ, AND MICHELE BESSO STRUGGLE WITH RELATIVITY

1. *Wikipedia*, s.v. "Time 100: The Most Important People of the Century," accessed December 11, 2015, https://en.wikipedia.org/wiki/Time_100:_The_Most_Important_People_of_the_Century.

2. Ibid.

3. Albert Einstein, Anna Beck, and Peter Havas, *The Collected Papers of Albert Einstein, Volume 1: The Early Years, 1879–1902* (Princeton, NJ: Princeton University Press, 1987), p. xviii.

4. Ibid.

5. Dennis Overbye, *Einstein in Love: A Scientific Romance* (New York: Penguin, 2000).

6. "Albert Einstein Quotes," BrainyQuote.com, accessed December 1, 2015, http://www.brainyquote.com/quotes/quotes/a/alberteins121678.html.

7. Jürgen Neffe, *Einstein: A Biography* (New York: Farrar, Straus, and Giroux, 2005), p. 31.

8. Ibid.

9. Jürgen Renn and Robert Schulmann, eds., *Albert Einstein/Mileva Maric: The Love Letters*, trans. Shawn Smith (Princeton, NJ: Princeton University Press, 1992), p. 41.

10. Ibid.

11. Ibid.

12. Overbye, *Einstein in Love*, p. 43.

13. "The Collected Papers of Albert Einstein; Volume 1, The Early Years, 1879–1902, no. 49," accessed September 2015, http://www.einsteinpapers.press.princeton.edu, http://einsteinpapers.press.princeton.edu/vol1-trans/150.

14. Overbye, *Einstein in Love*.

15. Barry R. Parker, *Creation: The Story of the Origin and Evolution of the Universe* (Berlin: Springer, 2013), p. 132.

16. Einstein, Beck, and Havas, *The Collected Papers of Albert Einstein, Volume 1*, p. xxii.

17. Karen C. Fox and Aries Keck, *Einstein A to Z* (New York: John Wiley & Sons, 2004), p. 19.

18. Albert Einstein, Anna Beck, and Peter Havas, *The Collected Papers of Albert Einstein, Volume 2: The Swiss Years: Writings, 1900–1909* (Princeton, NJ: Princeton University Press, 1989), p. 171.

CHAPTER 10: PART 2: ALBERT EINSTEIN'S STRUGGLES CONTINUE

1. Barry R. Parker, *Creation: The Story of the Origin and Evolution of the Universe* (Berlin: Springer, 2013), p. 172.

2. Walter Isaacson, *Einstein: His Life and Universe* (New York: Simon & Schuster, 2007).

3. Dennis Overbye, *Einstein in Love: A Scientific Romance* (New York: Penguin, 2000).

4. Galina Weinstein, "From the Berlin 'Entwurf' Field Equations to the Einstein Tensor I: October 1914 until Beginning of November 1915," p. 6, accessed January 7, 2016, http://arxiv.org/pdf/1201.5352.pdf.

5. Overbye, *Einstein in Love*, p. 252.

6. Ibid.

7. *Wikipedia*, s.v. "Emmy Noether," last modified January 1, 2016, https://en.wikipedia.org/wiki/Emmy_Noether.

8. Overbye, *Einstein in Love*.

9. Abraham Pais, *Subtle Is the Lord: The Science and the Life of Albert Einstein* (Oxford: Oxford University Press, 1982), p. 250.

10. "Albert Einstein and Zionism," Zionism & Israel Information Center, accessed July 2015, http://www.zionism-israel.com/Albert_Einstein/Albert_Einstein_zionism.htm.

11. Overbye, *Einstein in Love*, p. 353.

12. Manjit Kumar, *Quantum: Einstein, Bohr, and the Great Debate about the Nature of Reality* (New York: Norton, 2011).

13. Parker, *Creation*, p. 210.

14. Isaacson, *Einstein: His Life and Universe*.

15. *Wikiquote*, s.v. "Albert Einstein," last modified January 6, 2016, https://en .wikiquote.org/wiki/Albert_Einstein.

16. *Wikiquote*, s.v. "Niels Bohr," last modified January 1, 2016, https://en.wikiquote .org/wiki/Niels_Bohr.

17. Karen C. Fox and Aries Keck, *Einstein A to Z* (New York: John Wiley & Sons, 2004).

18. "Einstein's Quest for a Unified Theory," *APS News* 14, no. 11 (December 2005), accessed January 7, 2016, https://www.aps.org/publications/apsnews/200512/history.cfm.

19. Jürgen Neffe, *Einstein: A Biography* (New York: Farrar, Straus, and Giroux, 2005), p. 104.

20. Parker, *Creation*, p. 231.

21. "Albert Einstein in Caputh," Albert Einstein in the World Wide Web, accessed January 7, 2016, http://www.einstein-website.de/z_biography/caputh-e.html.

22. "Albert Einstein Quotes," ThinkExist.com, accessed January 7, 2016, http:// thinkexist.com/quotation/the_world_needs_heroes_and_it-s_better_they_be/15616.html.

23. Fox and Keck, *Einstein A to Z*, pp. 52–53.

24. Ibid., p. 129.

25. "Einstein's Quest for a Unified Theory."

CHAPTER 11: EDWIN HUBBLE AND HARLOW SHAPLEY CLASH/COOPERATE OVER THE UNIVERSE'S SIZE

1. *Wikiquote*, s.v. "J. B. S. Haldane," last modified December 19, 2015, https:// en.wikiquote.org/wiki/J._B._S._Haldane.

2. HubbleSite, accessed December 31, 2015, http://hubblesite.org.

3. Gale E. Christianson, *Edwin Hubble: Mariner of the Nebulae* (Chicago: University of Chicago Press, 1995), p. 25.

4. Ibid., p. 78.

5. Ibid., p. 82.

6. Ibid., p. 87.

7. Ibid., p. 109.

8. Ibid., p. 110.

9. SAO/NASA Astrophysics Data System, accessed January 7, 2016, http://adsabs .harvard.edu/full/1954MNRAS.114..291. P. 291.

10. Christianson, *Edwin Hubble*, p. 122.

11. Ibid.

12. Ibid., p. 132.

13. Carl C. Gaither and Alma E. Cavazos-Gaither, *Gaither's Dictionary of Scientific Quotations* (Berlin: Springer Science & Business Media, 2012), p. 200.

14. Christianson, *Edwin Hubble*, p. 158.

15. Edwin Hubble and Milton L. Humason, "The Velocity-Distance Relation among Extra-Galactic Nebulae," SAO/NASA Astrophysics Data System, accessed January 7, 2016, http://adsabs.harvard.edu/full/1931CMWCI.427....1H.

16. Ronald Clark, *Einstein: The Life and Times* (London: A&C Black, 2011).

17. Christianson, *Edwin Hubble*, p. 211.

18. *Wikipedia*, s.v. "Edwin Hubble," last modified December 29, 2015, https://en.wikipedia.org/wiki/Edwin_Hubble.

CHAPTER 12: DISASTROUS CONSEQUENCES OF LISE MEITNER AND OTTO HAHN'S DISCOVERY OF NUCLEAR FISSION

1. "Enrico Fermi," Radio Chemistry Society, accessed July 2015, http://www.radio chemistry.org/nuclearmedicine/pioneers/fermi_e.shtml.

2. *Wikipedia*, s.v. "Ida Noddack," last modified January 6, 2016, https://en.wikipedia .org/wiki/Ida_Noddack.

3. Laura Fermi, *Atoms in the Family: My Life with Enrico Fermi* (Chicago: University of Chicago Press, 2014), p. 139.

4. Dietrich Hahn, ed., *Otto Hahn—Leben und Werk in Texten und Bildern* (Frankfurt am Main, Germany: Suhrkamp Insel, 1988), p. 59.

5. Ruth Lewin Sime, *Lise Meitner, A Life in Physics* (Berkeley: University of California Press, 1996), p. 17.

6. Philipp Frank, *Einstein: His Life and Times* (London: Jonathan Cape, 1948), p. 139.

7. Sime, *Lise Meitner*, p. 233.

8. Lise Meitner and O. R. Frisch, "Products of the Fission of the Uranium Nucleus," *Nature* 143 (March 1939): 471–72.

9. "Niels Bohr Announces the Discovery of Fission," Atomic Heritage Foundation, January 28, 2015, accessed January 7, 2016, http://www.atomicheritage.org/article/ niels-bohr-announces-discovery-fission.

10. William Lanouette, *Genius in the Shadows: A Biography of Leo Szilard, the Man behind the Bomb* (New York: Skyhorse, 2015).

11. Cynthia C. Kelly, ed., *Manhattan Project: The Birth of the Atomic Bomb in the Words of Its Creators, Eyewitnesses, and Historians* (New York: Hachette, 2009), p. 39.

12. Lanouette, *Genius in the Shadows*.

13. *Wikipedia*, s.v. "Chicago Pile-1," last modified January 1, 2016, https://en.wikipedia .org/wiki/Chicago_Pile-1.

14. "Leo Szilard Facts," Your Dictionary, accessed January 7, 2016, http://biography .yourdictionary.com/leo-szilard.

15. John Amacher, "The Nazi Bomb: Failures of the German Nuclear Program," UCSB Oral History Project, June 2002, accessed January 7, 2016, http://www.history.ucsb.edu/ projects/holocaust/Research/Proseminar/johnamacher.htm.

16. *Wikipedia*, s.v. "Operation Epsilon," last modified December 1, 2015, https:// en.wikipcdia.org/wiki/Operation_Epsilon.

17. E. Amaldi, ed., "Reminiscences by Leo Szilard," in *Perspectives in American History* 2 (1968): 122–23.

18. Bernard T. Feld, "Einstein and the Politics of Nuclear Weapons," *Bulletin of the Atomic Scientists*, March 1979, p. 8.

19. Robert James Maddox, *Weapons for Victory: The Hiroshima Decision* (Columbia: University of Missouri Press, 2004), p. 67.

20. Kelly, *Manhattan Project*, p. 287.

21. Harry S. Truman and Robert H. Ferrell, *Off the Record: The Private Papers of Harry S. Truman* (Columbia: University of Missouri Press, 1997), p. 55.

22. Lawrence M. Krauss, "70 Years of Speaking Knowledge to Power," *Bulletin of the Atomic Scientists*, November 24, 2015, accessed January 7, 2016, http://thebulletin. org/70-years-speaking-knowledge-power8913.

23. Lansing Lamont, *Day of Trinity* (New York: Atheneum, 1985), pp. 332–33.

24. David B. Green, "This Day in Jewish History the Woman Who Discovered Nuclear Fission Dies," *Haaretz*, October 27, 2014, accessed January 7, 2016, http://www.haaretz .com/jewish/features/.premium-1.622862.

25. *Wikipedia*, s.v. "Lise Meitner," last modified January 7, 2016, https://en.wikipedia .org/wiki/Lise_Meitner.

26. *Wikipedia*, s.v. "Einstein-Szilárd Letter," last modified November 18, 2015, https:// en.wikipedia.org/wiki/Einstein%E2%80%93Szil%C3%A1rd_letter.

27. S. R. Weart and G. W. Szilárd, *Leo Szilard: His Version of the Facts* (Cambridge, MA: MIT Press, 1978).

28. William Lanouette, "The Many Worlds of Leo Szilard: Physicist, Peacemaker, Provocateur," SAO/NASA Astrophysics Data System, March 2014, accessed January 7, 2016, http://adsabs.harvard.edu/abs/2014APS..APRR17001L.

CHAPTER 13: MAURICE WILKINS, ROSALIND FRANKLIN, JAMES WATSON, AND FRANCIS CRICK DETERMINE THE STRUCTURE OF DNA

1. "Lewis Thomas Quotes," QuoteAddicts, accessed September 2015, http://quote addicts.com/844095.

2. *Wikipedia*, s.v. "Maurice Wilkins," last modified December 31, 2015, https://en.wikipedia.org/wiki/Maurice_Wilkins.

3. *Wikipedia*, s.v. "Francis Crick," last modified December 9, 2015, https://en.wikipedia.org/wiki/Francis_Crick.

4. James D. Watson, *A Passion for DNA Genes, Genomes, and Society* (Cold Spring Harbor, NY: Cold Spring Harbor Laboratory, 2000), p. 4.

5. Ibid.

6. J. D. Watson and F. H. C. Crick, "A Structure for Deoxyribose Nucleic Acid," *Nature* 171 (April 25, 1953): 171, 737–38, accessed January 7, 2016, http://www.exploratorium.edu/origins/coldspring/ideas/printit.html.

7. "The Nobel Prize in Physiology or Medicine 1962," Nobelprize.org, accessed January 7, 2016, http://www.nobelprize.org/nobel_prizes/medicine/laureates/1962/.

8. "Nobel Prizes 1962," Nobeprize.org, accessed January 7, 2016, http://www.nobelprize.org/nobel_prizes/lists/year/?year=1962.

9. *Wikipedia*, s.v. "Rosalind Franklin," last modified January 2, 2016, https://en.wikipedia.org/wiki/Rosalind_Franklin.

10. Anne Sayre, *Rosalind Franklin and DNA* (New York: W. W. Norton, 2000).

11. Jenifer Glynn, *My Sister Rosalind Franklin: A Family Memoir* (Oxford: Oxford University Press, 2012), p. 158.

12. "The Culture in Maurice Wilkins' Lab, Raymond Gosling," DNA Learning Center, accessed August 2015, https://www.dnalc.org/view/15261-The-culture-in-Maurice-Wilkins-lab- Raymond-Gosling.html.

13. James D. Watson, *The Double Helix: A Personal Account of the Discovery of the Structure of DNA* (New York: Touchstone, 1968).

14. Ira H. Carmen, *Politics in the Laboratory: The Constitution of Human Genomics* (Madison: University of Wisconsin Press, 2004), p. 31.

15. *Wikipedia*, s.v. "James Watson," last modified December 31, 2015, https://en.wikipedia.org/wiki/James_Watson.

16. Ibid.

17. "Russia's Usmanov to Give Back Watson's Auctioned Nobel Medal," BBC.com, December 9, 2014, accessed January 7, 2016, http://www.bbc.com/news/world-europe-30406322.

18. Keith Perry, "James Watson Selling Nobel Prize 'Because No-One Wants to Admit I Exist,'" *Telegraph*, November 28, 2014, accessed January 7, 2016, http://www.telegraph.co.uk/news/science/11261872/James-Watson-selling-Nobel-prize-because-no-one-wants-to-admit-I-exist.html.

CHAPTER 14: J. CRAIG VENTER, JAMES WATSON, AND MICHAEL HUNKAPILLER RACE FOR THE HUMAN GENOME

1. Robert M. Cook-Deegan, *The Gene Wars: Science, Politics, and the Human Genome* (New York: W. W. Norton, 1997), p. 88.

2. J. Craig Venter, *A Life Decoded: My Genome: My Life* (New York: Penguin, 2007), p. 110.

3. "A Brief History of the Human Genome Project," National Human Genome Research Institute, accessed January 7, 2016, http://www.genome.gov/12011239.

4. Ibid.

5. Kevin Davies, *Cracking the Genome: Inside the Race to Unlock Human DNA* (New York: Simon & Schuster, 2001), p. 60.

6. Ibid.

7. Venter, *A Life Decoded*, p. 137.

8. Cook-Deegan, *The Gene Wars*, p. 328.

9. Venter, *A Life Decoded*, p. 127.

10. Ibid., p. 173.

11. Shannon O'Lear, *Environmental Politics: Scale and Power* (Oxford: Oxford University Press, 2010), p. 55.

12. Venter, *A Life Decoded*, p. 207.

13. Ibid., p. 228.

14. Ed Regis, "Other People's Molecules," March 16, 2003, *New York Times*, accessed September 2015, http://www.nytimes.com/2003/03/16/books/other-people-s-molecules.html.

15. Venter, *A Life Decoded*, p. 278.

16. Davies, *Cracking the Genome*, p. 275.

17. Ibid., p. 236.

18. Ibid., p. 242.

CHAPTER 15: TEN HONORABLE MENTION MINI-CHAPTERS

1. Jennifer Latson, "The Shy Scientist Who Could See through Skin," *Time*, January 5, 2015, accessed September 2015, http://time.com/3649842/x-ray/.

2. *Wikipedia*, s.v. "Wilhelm Röntgen," last modified January 2, 2016, https://en.wikipedia.org/wiki/Wilhelm_R%C3%B6ntgen.

3. F. Zwicky, "Die Rotverschiebung von extragalaktischen Nebeln," *Helvetica Physica Acta* 6 (1933): 110.

4. *Wikipedia*, s.v. "Superconducting Super Collider," last modified January 3, 2016, https://en.wikipedia.org/wiki/Superconducting_Super_Collider.

5. "C. V. Raman Quotes," BrainyQuote.com, accessed September 2015, http://www .brainyquote.com/quotes/quotes/c/cvraman664315.html.

6. *Wikipedia*, s.v. "Satyendra Nath Bose," last modified January 3, 2016, https:// en.wikipedia.org/wiki/Satyendra_Nath_Bose.

7. "Early SETI: Project Ozma, Arecibo Message," SETI Institute, accessed January 7, 2016, http://www.seti.org/seti-institute/project/details/early-seti -project-ozma-arecibo-message.

8. "Stephen Hawking Quotes," BrainyQuote.com, accessed September 2015, http:// www.brainyquote.com/quotes/quotes/s/stephenhaw447580.html.

9. *Wikipedia*, s.v. "Stephen Hawking," last modified January 5, 2016, https:// en.wikipedia.org/wiki/Stephen_Hawking.

10. Ibid.

11. Ibid.

12. Ibid.

13. Steve Connor, "Higgs v Hawking: A Battle of the Heavyweights That Has Shaken the World of Theoretical Physics," *Independent*, September 2, 2002, accessed January 7, 2016, http://www.independent.co.uk/news/science/higgs-v-hawking-a-battle-of-the- heavyweights-that-has-shaken-the-world-of-theoretical-physics-175705.html.

14. *Wikipedia*, s.v. "Stephen Hawking."

15. Eliana Dockterman, "The True Story behind *The Theory of Everything*," *Time*, November 7, 2014, accessed January 7, 2016, http://time.com/3571702/theory -of-everything-true-story/.

16. *Wikipedia*, s.v. "Yuri Milner," last modified December 24, 2015, https:// en.wikipedia.org/wiki/Yuri_Milner.

17. *Wikipedia*, s.v. "Breakthrough Initiatives," last modified October 20, 2015, https:// en.wikipedia.org/wiki/Breakthrough_Initiatives.

18. Rachel Feltman, "Stephen Hawking Announces $100 Million Hunt for Alien Life," *Washington Post*, July 20, 2015, accessed January 7, 2016, https:// www.washingtonpost.com/news/speaking-of-science/wp/2015/07/20/stephen-hawking -announces-100-million-hunt-for-alien-life/.

19. "Carl Sagan Quotes," BrainyQuote.com, accessed September 2015, http://www .brainyquote.com/quotes/quotes/c/carlsagan133582.html.

20. "Ann Druyan Quotes," BrainyQuote.com, accessed September 2015, http://www .brainyquote.com/quotes/authors/a/ann_druyan.html.

21. "Neil deGrasse Tyson Quotes," BrainyQuote.com, accessed September 2015, http://www.brainyquote.com/quotes/quotes/n/neildegras615035.html.

22. Patrick Kevin Day, "Seth MacFarlane Donates Carl Sagan's Papers to Library of Congress," *Los Angeles Times*, June 28, 2012, accessed September 2015, http://articles.latimes.com/2012/jun/28/entertainment/la-et-st-seth-macfarlane-carl-sagan -library-congress-20120628.

23. Joel Achenbach, "Library of Congress Obtains Astronomer Carl Sagan's Personal

Papers," *Washington Post*, June 26, 2012, accessed January 7, 2016, https://www.washington post.com/national/health-science/library-of-congress-obtains-astronomer-carl-sagans -personal-papers/2012/06/26/gJQABdFN5V_story.html.

24. Carl Sagan and Ann Druyan, foreword to *The Demon-Haunted World: Science as a Candle in the Dark* (New York: Ballantine, 1996).

25. Joel Achenbach, "Why Carl Sagan Is Truly Irreplaceable," *Smithsonian*, March 2014, accessed August 2015, http://www.smithsonianmag.com/science-nature/ why-carl-sagan-truly-irreplaceable-180949818/.

26. "Carl Sagan Quotes," GoodReads, accessed January 6, 2016, http://www .goodreads.com/quotes/50379-extraordinary-claims-require-extraordinary-evidence.

27. Ruth Richman, "Lucky Ann Druyan Enjoys a Life of Curiosity," *Chicago Tribune*, November 15, 1992, accessed September 2015, http://articles.chicagotribune .com/1992-11-15/features/9204130523_1_ann-druyan-universe-space-aliens.

28. *Wikipedia*, s.v. "Carl Sagan," last modified, January 4, 2016, https://en.wikipedia. org/wiki/Carl_Sagan.

29. *Wikipedia*, s.v. "Neil deGrasse Tyson," last modified January 7, 2016, https:// en.wikipedia.org/wiki/Neil_deGrasse_Tyson.

30. "Neil deGrasse Tyson: Called by the Universe," Hayden Planetarium, accessed January 7, 2016, http://www.haydenplanetarium.org/tyson/read/2009/07/23/called -by-the-universe.

31. *Wikipedia*, s.v. "*Cosmos: A Spacetime Odyssey*," last modified December 30, 2015, https://en.wikipedia.org/wiki/Cosmos:_A_Spacetime_Odyssey.

32. "George Washington Carver Quotes," BrainyQuote.com, accessed September 2015, http://www.brainyquote.com/quotes/quotes/g/georgewash387003.html.

33. *Wikipedia*, s.v. "George Washington Carver," last modified January 4, 2016, https:// en.wikipedia.org/wiki/George_Washington_Carver.

34. *Wikipedia*, s.v. "Booker T. Washington," last modified December 30, 2015, https:// en.wikipedia.org/wiki/Booker_T._Washington.

35. Ibid.

36. Gary R. Kremer, *George Washington Carver: A Biography* (Santa Barbara, CA: ABC CLIO, 2011).

37. *Wikipedia*, s.v. "George Washington Carver."

38. "Hedy Lamarr Quotes," BrainyQuote.com, accessed September 2015, http://www .brainyquote.com/quotes/quotes/h/hedylamarr271434.html.

39. "Patents, Secret Communication System US 2292387 A," Google.com, accessed January 7, 2016, https://www.google.com/patents/US2292387.

40. Richard Rhodes, *Hedy's Folly: The Life and Breakthrough Inventions of Hedy Lamarr, the Most Beautiful Woman in the World* (New York: Vintage, 2011), p. 15.

41. "Patents, Secret Communication System US 2292387 A."

42. "Delilah Invented Secret Radio Control Torpedo System," *History Blog*, August 2011, accessed January 7, 2016, http://www.thehistoryblog.com/archives/date/2011/08/page/3.

43. "James Randi Quotes," BrainyQuote.com, accessed September 2015, http://www.brainyquote.com/quotes/quotes/j/jamesrandi395908.html.

44. *Wikipedia*, s.v. "James Randi," last modified December 20, 2015, https://en.wikipedia.org/wiki/James_Randi.

45. Adam Higginbotham, "The Unbelievable Skepticism of the Amazing Randi," *New York Times Magazine*, November 7, 2014, accessed August 2015, http://www.nytimes.com/2014/11/09/magazine/the-unbelievable-skepticism-of-the-amazing-randi.html?r=0.

46. James Randi Educational Foundation, accessed August 2015, http://web.randi.org/about.html.

AFTERWORD

1. *Wiktionary*, s.v. "The more things change, the more they stay the same," accessed January 7, 2016, https://en.wiktionary.org/wiki/the_more_things_change_the_more_they_stay_the_same.

TO DIG DEEPER

If you want to find out more about the people or ideas discussed here, consider the following sources:

GENERAL

Asimov, Isaac. *Asimov's Chronology of Science and Discovery*. New York: Harper & Row, 1989.

Ball, Philip. *Curiosity: How Science Became Interested in Everything*. Chicago and London: University of Chicago Press, 2012.

Bauer, Susan Wise. *The Story of Science*. New York: W. W. Norton, 2015.

Gribbin, John. *The Scientists: A History of Science Told through the Lives of Its Greatest Inventors*. New York: Random House, 2002.

Hellman, Hal. *Great Feuds in Science: Ten of the Liveliest Disputes Ever*. New York: John Wiley & Sons, 1998.

Levy, Joel. *Scientific Feuds: From Galileo to the Human Genome Project*. London: New Holland, 2010.

Livio, Mario. *Brilliant Blunders*. New York: Simon & Schuster, 2013.

Myers, Morton A. *Prize Fight: The Race and the Rivalry to be the First in Science*. New York: Palgrave Macmillan, 2012.

Robinson, Matthew, ed. *The Scientists: An Epic of Discovery*. New York: Thames and Hudson, 2012.

Simmons, John. *The Scientific 100: A Ranking of the Most Influential Scientists, Past and Present*. Secaucus, NJ: Citadel, 1996.

INTRODUCTION

"Scientific Method." *Wikipedia*, s.v. Last modified December 6, 2015. http://en.wikipedia.org/wiki/Scientific_method.

"Understanding and Using the Scientific Method." Science Made Simple. Accessed May 2015. http://www.sciencemadesimple.com/scientific_method.html.

Wynn Sr., Charles M., and Arthur W. Wiggins. *The Five Biggest Ideas in Science*. New York: Wiley, 1997.

CHAPTER 1: DEMOCRITUS AND ARISTOTLE PONDER THE EXISTENCE OF ATOMS

"Aristotle." *Wikipedia*, s.v. Last modified January 4, 2016. http://en.wikipedia.org/wiki/Aristotle.

"Aristotle (384–322 BCE)." In *Internet Encyclopedia of Philosophy*. Accessed May 2015. http://www.iep.utm.edu/aristotl/.

Berryman, Sylvia. "Democritus." In *Stanford Encyclopedia of Philosophy*, edited by Edward N. Zalta, 2010. Accessed May 2015. http://plato.stanford.edu/entries/democritus/.

"Democritus." European Graduate School. Accessed May 2015. http://www.egs.edu/library/democritus/biography/.

"Democritus." *Wikipedia*, s.v. Last modified December 2, 2015. http://en.wikipedia.org/wiki/Democritus.

Shields, Christopher. "Aristotle." In *The Stanford Encyclopedia of Philosophy*, edited by Edward N. Zalta, 2010. Accessed May 2015. http://plato.stanford.edu/archives/fall2015/entries/aristotle.

Wynn Sr., Charles M., and Arthur W. Wiggins. *The Five Biggest Ideas in Science*. New York: Wiley, 1997.

Bonus video interview: "Atoms with Democritus and Aristotle." Vimeo video, 18:01. Posted on June 2015 by Bloomfield Township. Accessed January 4, 2016. http://vimeo.com/125063018.

CHAPTER 2: ARISTOTLE, ARISTARCHUS, COPERNICUS, AND GALILEO SEEK TO DETERMINE EARTH'S PLACE IN THE COSMOS

Gannon, Megan. "Tycho Brahe Died from Pee, Not Poison." *Live Science*, November 6, 2012. Accessed May 2015. http://www.livescience.com/24835-astronomer-tycho-brahe-death.html.

"Ptolemy." *Wikipedia*, s.v. Last modified January 5, 2016. http://en.wikipedia.org/wiki/Ptolemy.

Rubenstein. Richard E. *Aristotle's Children: How Christians, Muslims, and Jews Rediscovered Ancient Wisdom and Illuminated the Dark Ages*. New York: Harcourt, 2004.

Wiggins, Arthur W., and Charles M. Wynn Sr. *The Five Biggest Unsolved Problems in Science*. New York: Wiley, 2003.

Bonus Video Interview: "The Solar System with Tycho Brahe and Galileo Galilei." Vimeo video, 14:45. Posted on July 2015 by Bloomfield Township. http://vimeo.com/129116467.

CHAPTER 3: ISAAC NEWTON, ROBERT HOOKE, AND GOTTFRIED LEIBNIZ ARGUE ABOUT MOTION AND CALCULUS

Clark, David H., and Stephen P. H. Clark. *Newton's Tyranny: The Suppressed Scientific Discoveries of Stephen Gray and John Flamsteed.* New York: W. H. Freeman, 2001.
"Edmond Halley." *Wikipedia*, s.v. Last modified January 5, 2016. http://en.wikipedia.org/wiki/Edmond_Halley.
"Gottfried Wilhelm Leibniz." *Wikipedia*, s.v. Last modified January 4, 2016. http://en.wikipedia.org/wiki/Gottfried_Wilhelm_Leibniz.
"Isaac Newton." *Wikipedia*, s.v. Last modified January 4, 2016. http://en.wikipedia.org/wiki/Isaac_Newton.
"Nicolas Fatio de Duillier." *Wikipedia*, s.v. Last modified June 1, 2015. http://en.wikipedia.org/wiki/Nicolas_Fatio_de_Duillier.
McNab, Andrew. "On the Shoulders of Giants." IsaacNewton.org.uk. Accessed September 2015. http://www.isaacnewton.org.uk/essays/Giants.
"Robert Hooke." *Wikipedia*, s.v. Last modified November 16, 2015. http://en.wikipedia.org/wiki/Robert_Hooke.
Wiggins, Arthur W. *The Joy of Physics.* Amherst, NY: Prometheus Books, 2007; 2nd ed., 2011.
Bonus Video Interview: "Newton and Hooke on Gravity." Vimeo video, 14:47. Posted on July 2015 by Bloomfield Township. http://vimeo.com/130318251.

CHAPTER 4: THE BATTLING BERNOULLIS AND BERNOULLI'S PRINCIPLE

Bui, Dung (Yom), and Mohamed Allali. "The Bernoulli Family: Their Massive Contributions to Mathematics and Hostility toward Each Other." *Academia* 2, no. 2 (2011). Accessed May 2015. http://www.academia.edu/6645678/The_Bernoulli_Family_their_massive_contributions_to_mathematics_and_hostility_toward_each_other.
"18th Century Mathematics—Bernoulli Brothers." storyofmathematics.com. Accessed May 2015. http://www.storyofmathematics.com/18th_bernoulli.html.
Gonzales, Tina. "Family Squabbles: The Bernoulli Family." math.wichita.edu. Accessed May 2015. http://www.math.wichita.edu/history/Men/bernoulli.html.
"Guillaume de l'Hôpital." *Wikipedia*, s.v. Last modified November 18, 2015. http://en.wikipedia.org/wiki/Guillaume_de_l%27Hôpital.
"Later Life of Isaac Newton." *Wikipedia*, s.v. Last modified December 16, 2015. http://en.wikipedia.org/wiki/Later_life_of_Isaac_Newton.

O'Connor, J. J., and E. F. Robertson. "Jacob (Jacques) Bernoulli." MacTutor History of Mathematics Archive. Accessed May 2015. http://www-history.mcs.st-and .ac.uk/Biographies/Bernoulli_Jacob.html.

———. "The Brachistochrone Problem." MacTutor History of Mathematics Archive. Accessed May 2015. http://www-history.mcs.st- and.ac.uk/HistTopics/Brachisto chrone.html.

CHAPTER 5: ANTOINE LAVOISIER AND BENJAMIN THOMPSON (COUNT RUMFORD) HAVE RIVAL THEORIES OF HEAT

"Benjamin Thompson." Famous Scientists. Accessed June 2015. http://www.famous scientists.org/benjamin-thompson/.

Brown, G. I. *Count Rumford: The Extraordinary Life of a Scientific Genius*. Glouces- tershire, UK: Sutton, 1999.

Brown, Sanborn C. "Count Rumford and the Caloric Theory of Heat." *Proceedings of the American Philosophical Society* 93, no. 4 (1949): 316–25. Accessed June 2015. http://www.jstor.org/discover/10.2307/3143157?sid=21105819778481&uid =70&uid=4&uid=2129&uid=3739600&uid=3739256&uid=2.

"Caloric Theory." *Wikipedia*, s.v. Last modified December 25, 2015. http://en .wikipedia.org/wiki/Caloric_theory.

"Cannon Boring Experiment." In *Encyclopedia of Human Thermodynamics*. Accessed June 2015. http://www.eoht.info/page/Cannon+boring+experiment.

"Charles Theodore, Elector of Bavaria." *Wikipedia*, s.v. Last modified November 19, 2015. http://en.wikipedia.org/wiki/Charles_Theodore,_Elector_of_Bavaria.

"The Chemical Revolution of Antoine-Laurent Lavoisier." Historic Chemical Land- marks program of the American Chemical Society International. Accessed June 2015. http://www.acs.org/content/acs/en/education/whatischemistry/landmarks/ lavoisier.html.

Gurstelle, William. *The Practical Pyromaniac*. Chicago:

Hoffmann, Roald. "Mme. Lavoisier." *American Scientist*. Accessed June 2015. http:// www.americanscientist.org/issues/pub/mme-lavoisier.□□□□□□□□□

Lienhard, John H. "Marie Lavoisier." In *Engines of Our Ingenuity*, University of Houston. Accessed June 2015. http://www.uh.edu/engines/epi1673.htm.□

"Marie-Anne Paulze Lavoisier." *Wikipedia*, s.v. Last modified December 1, 2015. http://en.wikipedia.org/wiki/Marie-Anne_Paulze_Lavoisier.

Morris, Robert J. "Lavoisier and the Caloric Theory." NYU Tandon School of Engineering. Accessed June 2015. http://www.faculty.poly.edu/~jbain/heat/ readings/72Morris.pdf.

Rumford, Benjamin Count of. "An Inquiry concerning the Source of the Heat Which Is Excited by Friction." *Philosophical Transactions of the Royal Society of London* 88 (1798): 80–102. Accessed June 2015. http://www.jstor.org/stable/106970?seq=7#page_scan_tab_contents.

"Sir Benjamin, Thompson, Count Rumford (1753–1814)." Burgum Family History Society. freepages.genealogy.rootsweb.ancestry.com. Accessed June 2015. http://freepages.genealogy.rootsweb.ancestry.com/~bfhs/chap2.html.

Wiggins, Arthur W. *The Joy of Physics*. Amherst, NY: Prometheus Books, 2007; 2nd ed., 2011.

Bonus Video Interview: "Lavoisier and Thompson on Heat." Vimeo video, 11:55. Posted on August 2015 by Bloomfield Township. http://vimeo.com/132827829.

CHAPTER 6: MENDELEEV, MEYER, MOSELEY AND THE BIRTH OF THE PERIODIC TABLE

Aldersey-Williams, Hugh. *Periodic Tales: A Cultural History of the Elements from Arsenic to Zinc*. New York: HarperCollins, 2011.

"A Brief History of the Development of Periodic Table." Western Oregon University. Accessed June 2015. https://www.wou.edu/las/physci/ch412/perhist.htm.

"Dmitri Mendeleev." Famous Scientists. Accessed June 2015. http://www.famous scientists.org/dmitri-mendeleev/.

Gordin, Mishael D. "The Textbook Case of a Priority Dispute: D. I. Mendeleev, Lothar Meyer and the Periodic System." *From Nature Engaged; Science in Practice from the Renaissance to the Present*, edited by Mario Biagioli and Jessica Riskin. London: Palgrave Macmillan, 2012. Accessed June 2015.

"History of the Periodic Table." *Wikipedia*, s.v. Last modified January 4, 2016. http://en.wikipedia.org/wiki/History_of_the_periodic_table.

"Julius Lothar Meyer." *Wikipedia*, s.v. Last modified January 5, 2016. http://en.wikipedia.org/wiki/Julius_Lothar_Meyer.

"Julius Lothar Meyer and Dmitri Ivanovich Mendeleev." Chemical Heritage Foundation. Accessed June 2015. http://www.chemheritage.org/discover/online-resources/chemistry-in-history/themes/the-path-to-the-periodic-table/meyer-and-mendeleev.aspx.

Kean, Sam. *The Disappearing Spoon: And Other True Tales of Madness, Love, and the History of the World from the Periodic Table of the Elements*. New York: Little Brown, 2010.

"Periodic Table." Royal Society of Chemistry. Accessed June 2015. http://www.rsc.org/periodic-table/history/about.

Van der Krogt, Peter. "Development of the Chemical Symbols and the Periodic Table."

VanderKrogt.net. Accessed June 2015. http://www.vanderkrogt.net/elements/chemical_symbols.php.

Wynn Sr., Charles M., and Arthur W. Wiggins. *The Five Biggest Ideas in Science*. New York: Wiley, 1997.

CHAPTER 7: WESTINGHOUSE AND TESLA VERSUS EDISON — AC/DC TITANS CLASH

Cheney, Margaret. *Tesla: Man Out of Time*. New York: Barnes & Noble, 1981.

Clark, Ronald. *Edison: The Man Who Made the Future*. London: A&C Black, 2012.

Kent, David J. *Tesla: The Wizard of Electricity*. New York: Fall River, 2015.

"Lighting the 1893 World's Fair: The Race to Light the World." *History Rat*. Accessed January 5, 2016. https://historyrat.wordpress.com/2013/01/13/lighting-the-1893-worlds-fair-the-race-to-light-the-world/.

Santoso, Alex. "10 Fascinating Facts about Edison." *Neatorama*. February 11, 2008. Accessed January 5, 2016. http://www.neatorama.com/2008/02/11/10-fascinating-facts-about-edison/.

Seifer, Mark J. *The Life and Times of Nikola Tesla: Biography of a Genius*. New York: Citadel, 1998.

Stewart, Daniel Blair. *Tesla: The Modern Sorcerer*. Bombay, India: Frog Books, 1999.

Valone, Thomas. *Harnessing the Wheelwork of Nature: Tesla's Science of Energy*.

Wiggins, Arthur W. *The Joy of Physics*. Amherst, NY: Prometheus Books, 2007; 2nd ed., 2011.

Bonus Video Interview: "Tesla-Edison Clash over AC/DC." Vimeo video, 11:54. Posted on September 2015 by Bloomfield Township. http://vimeo.com/135971779.

CHAPTER 8: ALFRED WEGENER STANDS HIS GROUND ABOUT CONTINENTAL DRIFT

"Alfred Wegener." *Wikipedia*, s.v. Last modified November 13, 2015. http://en.wikipedia.org/wiki/Alfred_Wegener.

"Alfred Wegener (1880–1930)." University of California Museum of Paleontology. Accessed June 2015. http://www.ucmp.berkeley.edu/history/wegener.html.

Maugh II, Thomas H. "Victor Vacquier Sr. Dies at 101; Geophysicist Was a Master of Magnetics." *Los Angeles Times*. January 24, 2009. Accessed January 5, 2016. http://www.latimes.com/science/la-me-vacquier24-2009jan24-story.html.

Sant, Joseph. "Wegener and Continental Drift Theory." Scientus.org. Accessed June 2015. http://www.scientus.org/Wegener-Continental-Drift.html.

"Victor Vacquier." *Wikipedia*, s.v. Last modified June 5, 2015. http://en.wikipedia.org/wiki/Victor_Vacquier.

Wynn Sr., Charles M., and Arthur W. Wiggins. *The Five Biggest Ideas in Science*. New York: Wiley, 1997.

CHAPTER 9: PART 1: ALBERT EINSTEIN, MARCEL GROSSMANN, MILEVA MARIĆ, AND MICHELE BESSO STRUGGLE WITH RELATIVITY

"Einstein Family." *Wikipedia*, s.v. Last modified January 4, 2016. http://en.wikipedia.org/wiki/Einstein_family.

"Heinrich Friedrich Weber." *Wikipedia*, s.v. Last modified April 30, 2015. http://en.wikipedia.org/wiki/Heinrich_Friedrich_Weber.

"Heinrich Zangger." *Wikipedia*, s.v. Last modified August 12, 2015. https://de.wikipedia.org/wiki/Heinrich_Zangger.

Lanouette, William, with Bela Silard. *Genius in the Shadows: A Biography of Leo Szilard, the Man behind the Bomb*. New York: Charles Scribner's Sons, 1992.

"Lenard, Philipp." In *Complete Dictionary of Scientific Biography*. Charles Scribner's Sons, 2008. Encyclopedia.com. Accessed July 2015. http://www.encyclopedia.com/topic/Philipp_Lenard.aspx.

"Marcel Grossmann." In *Complete Dictionary of Scientific Biography*. Charles Scribner's Sons, 2008. Encyclopedia.com. Accessed July 2015. http://www.encyclopedia.com/topic/Marcel_Grossmann.aspx.

"Marcel Grossmann." *Wikipedia*, s.v. Last modified November 24, 2015. http://en.wikipedia.org/wiki/Marcel_Grossmann.

"Mileva Marić." *Wikipedia*, s.v. Last modified November 27, 2015. http://en.wikipedia.org/wiki/Mileva_Marić.

Neffe, Jürgen. *Einstein: A Biography*. New York: Farrar, Straus, and Giroux, 2005.

Overbye, Dennis. *Einstein in Love: A Scientific Romance*. New York: Penguin, 2000.

Parker, Barry. *Einstein: The Passions of a Scientist*. Amherst, NY: Prometheus Books, 2003.

Seelig, Carl. *Albert Einstein: A Documentary Biography*. Zurich: Staples, 1956.

"Time 100: The Most Influential People of the Century." *Wikipedia*, s.v. Last modified December 11, 2015. http://en.wikipedia.org/wiki/Time_100:_The_Most_Important_People_of_the_Century.

Weinstein, Galina. *Einstein's Pathway to the Special Theory of Relativity*. Newcastle upon Tyne, UK: Cambridge Scholars, 2015.

Wiggins, Arthur W. *The Joy of Physics*. Amherst, NY: Prometheus Books, 2007; 2nd ed., 2011.

CHAPTER 10: PART 2: ALBERT EINSTEIN'S STRUGGLES CONTINUE

"Albert Einstein and Zionism." Zionism & Israel Information Center. Accessed July 2015. http://www.zionism-israel.com/Albert_Einstein/Albert_Einstein_zionism.htm.

Berlinski, David. "Einstein and Gödel." *Discover*, March 1, 2002. Accessed July 2015. http://discovermagazine.com/2002/mar/featgodel.

Einstein, Albert, with Paul A. Schilpp, transl. and ed. *Albert Einstein: Autobiographical Notes*. LaSalle, IL, and Chicago: Open Court, 1979.

"Einstein's Quest for a Unified Theory." *APS News* 14, no. 11 (December 2005), American Physical Society. Accessed July 2015. http://www.aps.org/publications/apsnews/200512/history.cfm.

Gewertz, Ken. "Albert Einstein, Civil Rights Activist." *Harvard Gazette*, April 12, 2007. Accessed July 2015. http://news.harvard.edu/gazette/story/2007/04/albert-einstein-civil-rights-activist/.

Hillman, Bruce J., Birgit Ertl-Wagner, and Bernd C. Wagner. *The Man Who Stalked Einstein: How Nazi Scientist Phillip Lenard Changed the Course of History*. Guildford, CT: Lyons, 2015.

Isaacson, Walter. "How Einstein Divided America's Jews." *Atlantic*, December 2009. Accessed July 2015. http://www.theatlantic.com/magazine/archive/2009/12/how-einstein-divided-americas-jews/307763/.

Jerome, Fred, and Rodger Taylor. *Einstein on Race and Racism*. New Brunswick, NJ: Rutgers University Press, 2006.

Katz, William Loren. "Albert Einstein, Paul Robeson, and Israel." *William Loren Katz*, January 21, 2006. Accessed July 2015. http://williamlkatz.com/einstein-robeson-israel/.

Kumar, Manjit. *Quantum: Einstein, Bohr, and the Great Debate about the Nature of Reality*. New York: Norton, 2011.

Levenson, Thomas. *Einstein in Berlin*. New York: Bantam, 2003.

"May 3, 1946—Albert Einstein Spoke at Lincoln University in Pennsylvania." *Rhapsody in Books*, May 3, 2010. Accessed July 2015. https://rhapsodyinbooks.wordpress.com/2010/05/03/may-3-1946-%E2%80%93-albert-einstein-spoke-at-lincoln-university-in-pennsylvania/.

"Offering the Presidency of Israel to Albert Einstein." Jewish Virtual Library. Accessed July 2015. https://www.jewishvirtuallibrary.org/jsource/Politics/einsteinlet.html.

Pirro, Deirdre. "Maria (Maja) Einstein: A Tuscan Paradise Lost." *Florentine*, no. 207 (February 2015). Accessed July 2015. http://www.theflorentine.net/articles/article-view.asp?issuetocId=9814.

Popova, Maria. "Albert Einstein's Little-Known Correspondence with W. E. B. Du Bois on Race and Racial Justice." *Brain Pickings*. Accessed July 2015. https://www.brainpickings.org/2015/01/06/albert-einstein-w-e-b-du-bois-racism/.

Rosenkranz, Ze'ev. *Einstein before Israel: Zionist Icon or Iconoclast?* Princeton, NJ: Princeton University Press, 2011.

Sayen, Jamie. *Einstein in America: The Scientist's Conscience in the Age of Hitler and Hiroshima.* New York: Crown, 1985.

Singer, Saul Jay. "Weizmann and Einstein: The Succession That Wasn't." *Jewish Press*, May 28, 2015. Accessed July 2015. http://www.jewishpress.com/indepth/front-page/weizmann-and-einstein-the-succession-that-wasnt/2015/05/28/.

Wiggins, Arthur W. *The Joy of Physics.* Amherst, NY: Prometheus Books, 2007; 2nd ed., 2011.

Yourgrau, Palle. *A World without Time: The Forgotten Legacy of Godel and Einstein.*

Bonus Video Interview: "Bohr and Einstein Differ on the Quantum." Vimeo video, 19:47. Posted on October 2015 by Bloomfield Township. http://vimeo.com/138738811.

CHAPTER 11: EDWIN HUBBLE AND HARLOW SHAPLEY CLASH/COOPERATE OVER THE UNIVERSE'S SIZE

Christianson, Gale E. *Edwin Hubble: Mariner of the Nebulae.* Chicago: University of Chicago Press, 1995.

"Edwin Powell Hubble—The Man Who Discovered the Cosmos." Hubble Space Telescope. Accessed July 2015. https://www.spacetelescope.org/about/history/the_man_behind_the_name/.

Fernie, J. D. "The Period-Luminosity Relation: A Historical Review." *Publications of the Astronomical Society of the Pacific* 81, no. 483:707. Accessed July 2015. http://adsabs.harvard.edu/full/1969PASP...81..707F.

"From Our Galaxy to Island Universes." Ideas of Cosmology. Accessed July 2015. https://www.aip.org/history/cosmology/ideas/island.htm.

Geiling, Natasha. "The Women Who Mapped the Universe and Still Couldn't Get Any Respect." Smithsonian.com, September 18, 2003. Accessed July 2015. http://www.smithsonianmag.com/history/the-women-who-mapped-the-universe-and-still-couldnt-get-any-respect-9287444/.

Glass, Ian S. *Revolutionaries of the Cosmos: The Astro-Physicists.* London:

"Henrietta Leavitt 1868–1921." PBS.org. Accessed July 2015. http://www.pbs.org/wgbh/aso/databank/entries/baleav.html.

"Henry Draper (1837–1882)." Open Door Web Site. History of Science and Technology. Accessed July 2015. http://www.saburchill.com/HOS/astronomy/033.html.

Johnson, George. *Miss Leavitt's Stars.* New York: W. W. Norton, 2005.

Levy, David. *The Scientific American Book of the Cosmos.* New York: St. Martin's, 2000.

"Milton L. Humason." *Wikipedia*, s.v. Last modified May 4, 2015. https://en.wikipedia
.org/wiki/Milton_L._Humason.

North, John. *Cosmos: An Illustrated History of Astronomy and Cosmology*. Chicago:

Parker, Barry R. *Creation: The Story of the Origin and Evolution of the Universe*.
Berlin: Springer 2013.

Putnam, William Lowell, et al. *The Explorers of Mars Hill*. Flagstaff, AZ: Lowell
Observatory, 1994.

"Shapley, Harlow." In *Complete Dictionary of Scientific Biography*, 2008. Encyclo-
pedia.com. Accessed July 2015. http://www.encyclopedia.com/topic/Harlow
_Shapley.aspx.

Shapley, Harlow. *Through Rugged Ways to the Stars*. New York: Scribner's Sons,
1969.

"Slipher, Vesto Melvin." In *Complete Dictionary of Scientific Biography*, 2008.
Encyclopedia.com, December 11, 2015. Accessed July 2015. http://www
.encyclopedia.com/topic/Vesto_Melvin_Slipher.aspx#2.

Wiggins, Arthur W., and Charles M. Wynn Sr. *The Five Biggest Unsolved Problems in
Science*. New York: Wiley, 2003.

Bonus Video Interview: "Hubble and Shapley Figure the Universe Size." Vimeo
video, 16:14. Posted on September 2015 by Bloomfield Township. http://vimeo
.com/135583736.

CHAPTER 12: DISASTROUS CONSEQUENCES OF LISE MEITNER AND OTTO HAHN'S DISCOVERY OF NUCLEAR FISSION

Barron, Rachel Stiffler. *Lise Meitner, Discoverer of Nuclear Fission*. Greensboro, NC:
Morgan Reynolds, 2000.

"The Bohr-Heisenberg Meeting in September 1941." American Institute of Physics.
Accessed July 2015. https://www.aip.org/history/heisenberg/bohr-heisenberg
-meeting.htm.

Bortz, Alfred B. *Physics: Decade by Decade*. New York: Facts on File, 2007.

Bulletin of the Atomic Scientists. Accessed July 2015. http://thebulletin.org.

"Discovery of Nuclear Fission." *APS News* 16, no. 11 (December 2007), American Physical Society. Accessed July 2015. http://www.aps.org/publications/
apsnews/200712/physicshistory.cfm.

"Enrico Fermi." Radio Chemistry Society. Accessed July 2015. http://www.radio
chemistry.org/nuclearmedicine/pioneers/fermi_e.shtml.

Frisch, Otto. *What Little I Remember*. Cambridge: Cambridge University Press, 1979.

Fromm, James Richard. *Harnessing of Nuclear Fission: The Story of the Atomic*

Bomb. Accessed July 2015. Third Millennium Online. http://www.3rd1000.com/ nuclear/cruc18.htm.

"German Nuclear Weapon Project." *Wikipedia*, s.v. Last modified December 29, 2015. https://en.wikipedia.org/wiki/German_nuclear_weapon_project.

Hafemeister, David, ed. *Physics and Nuclear Arms Today*. Melville, NY: American Institute of Physics, 1991.

"James Chadwick." *Wikipedia*, s.v. Last modified December 16, 2015. https:// en.wikipedia.org/wiki/James_Chadwick.

Jungk, Robert. *Brighter Than a Thousand Suns*. Boston: Mariner, 1970.

Lanouette, William. "Ideas by Szilard, Physics by Fermi." *Bulletin of the Atomic Scientists* 48, no. 10 (1992): 16–23.

"Manhattan Project." *Wikipedia*, s.v. Last modified January 5, 2016. https://en .wikipedia.org/wiki/Manhattan_Project.

"The Manhattan Project: Making the Atomic Bomb." Atomicarchive.com. Accessed July 2015. http://www.atomicarchive.com/History/mp/.

"May 1932: Chadwick Reports the Discovery of the Neutron." *APS News* 16, no. 5 (May 2007), American Physical Society. Accessed July 2015. http://www.aps .org/publications/apsnews/200705/physicshistory.cfm.

Rhodes, Richard. *The Making of the Atomic Bomb*. New York: Simon & Schuster, 1986.

Rife, Patricia. "Lise Meitner." In *Jewish Women: A Comprehensive Historical Encyclopedia* (2009). Jewish Women's Archive. Accessed July 2015. http://jwa.org/ encyclopedia/article/meitner-lise.

Segré, Emilio. *Enrico Fermi, Physicist*. Chicago: University of Chicago Press, 1970.

Several, James Lewis. *World War II*. Google Play: Kreactive Editorial, n.d.

Sime, Ruth Lewin. *Lise Meitner, A Life in Physics*. Berkeley: University of California Press, 1996.

"Werner Heisenberg." *Wikipedia*, s.v. Last modified December 31, 2015. http:// en.wikipedia.org/wiki/Werner_Heisenberg.

Wiggins, Arthur W. *The Joy of Physics*. Amherst, NY: Prometheus Books, 2007; 2nd ed., 2011.

CHAPTER 13: MAURICE WILKINS, ROSALIND FRANKLIN, JAMES WATSON, AND FRANCIS CRICK DETERMINE THE STRUCTURE OF DNA

Borell, Brendan. "Watson's Nobel Medal Sells for US$4.1 Million." *Nature* (December 4, 2014). Accessed July 2015. http://www.nature.com/news/ watson-s-nobel-medal- sells-for-us-4-1-million-1.16500.

Crick, Francis. *What Mad Pursuit; A Personal View of Scientific Discovery*. New York: Basic, 1990.

"The Culture in Maurice Wilkins' Lab, Raymond Gosling." DNA Learning Center. Accessed August 2015. https://www.dnalc.org/view/15261-The-culture-in-Maurice-Wilkins-lab- Raymond-Gosling.html.

"DNA Story at King's: The Hidden DNA Workers." *DNA and Social Responsibility*. Accessed August 2015. http://dnaandsocialresponsibility.blogspot.com/2010_09_01_archive.html.

"Francis Crick (1916–2004)." DNA from the Beginning. Accessed August 2015. http://www.dnaftb.org/19/bio-2.html.

Hall, Kersten T. *The Man in the Monkeynut Coat: William Astbury and the Forgotten Road to the Double-Helix*. Oxford: Oxford University Press, August 2014.

Maddox, Brenda. *Rosalind Franklin: The Dark Lady of DNA*. New York: Harper-Collins, 2002.

Perry, Keith. "James Watson Selling Nobel Prize 'Because No-One Wants to Admit I Exist.'" *Telegraph*, November 28, 2014. Accessed August 2015. http://www.telegraph.co.uk/news/science/11261872/James-Watson-selling-Nobel-prize-because-no-one-wants-to-admit-I-exist.html.

Piper, Anne. "Rosalind Franklin." In *Trends in Biochemical Sciences* 23 (1998): 151–54. Accessed August 2015. http://cwp.library.ucla.edu/articles/franklin/piper.html#section9.

"Raymond Gosling on Working on DNA with Rosalind Franklin." Cold Spring Harbor Laboratory Oral History Collection, recorded March 3, 2003. Accessed January 7, 2016. http://library.cshl.edu/oralhistory/interview/scientific-experience/molecular-biologists/working-dna-rosalind-franklin/.

Ridley, Matt. *Francis Crick: Discoverer of the Genetic Code*. New York: Harper, 2006.

Sayre, Anne. *Rosalind Franklin and DNA*. New York: W. W. Norton, 2000.

Watson, James D. *A Passion for DNA Genes, Genomes, and Society*. Cold Spring Harbor, NY: Cold Spring Harbor Laboratory, 2000.

Watson, James D. *The Double Helix: A Personal Account of the Discovery of the Structure of DNA*. New York: Touchstone, 1968.

Wilkins, Maurice. *The Third Man of the Double Helix: The Autobiography of Maurice Wilkins*. Oxford: Oxford University Press, 2003.

"William Astbury." *Wikipedia*, s.v. Last modified January 5, 2016. https://en.wikipedia.org/wiki/William_Astbury.

Bonus Video Interview: "Franklin and Wilkins Do DNA." Vimeo video, 11:49. Posted on October 2015 by Bloomfield Township. http://vimeo.com/138738821.

CHAPTER 14: J. CRAIG VENTER, JAMES WATSON, AND MICHAEL HUNKAPILLER RACE FOR THE HUMAN GENOME

Cook-Deegan. Robert. *The Gene Wars: Science, Politics, and the Human Genome.* New York: W. W. Norton, 1998.

Davies, Kevin. *Cracking the Genome: Inside the Race to Unlock Human DNA.* New York: Simon & Schuster, 2001.

"Genome Sizes." users.rcn.com. Accessed August 2015. http://users.rcn.com/jkimball.ma.ultranet/BiologyPages/G/GenomeSizes.html.

Hood, Lee. "Lee Hood: Making the Future Happen." *Scientific American Worldview.* Accessed August 2015. http://www.saworldview.com/archive/2012/lee-hood-making-the-future-happen/.

"Leroy Hood." *Wikipedia*, s. v. Last modified November 22, 2015. https://en.wikipedia.org/wiki/Leroy_Hood.

Ridley, Matt. *Genome: The Autobiography of a Species in 23 Chapters.* New York: Harper, 2000.

Venter, J. Craig. *A Life Decoded: My Genome: My Life.* New York: Viking, 2007.

Wiggins, Arthur W., and Charles M. Wynn Sr. *The Five Biggest Unsolved Problems in Science.* New York: Wiley, 2003.

CHAPTER 15.1: TEN HONORABLE MENTIONS: WILHELM CONRAD RÖNTGEN

Berger, Harold. *The Mystery of a New Kind of Rays: The Story of Wilhelm Conrad Roentgen and His Discovery of X-Rays.* CreateSpace, 2012.

Markel, Howard. "'I Have Seen My Death': How the World Discovered the X-Ray." PBS NewsHour, December 20, 2012. Accessed August 2015. http://www.pbs.org/newshour/rundown/i-have-seen-my-death-how-the-world-discovered-the-x-ray/.

"Wilhelm Conrad Röntgen—Biographical." Nobelprize.org. Nobel Media AB 2014. Accessed August 2015. http://www.nobelprize.org/nobel_prizes/physics/laureates/1901/rontgen-bio.html.

"Wilhelm Röntgen." *Wikipedia*, s.v. Last modified January 2, 2016. https://en.wikipedia.org/wiki/Wilhelm_Röntgen.

CHAPTER 15.2: TEN HONORABLE MENTIONS: PERLMUTTER, RIESS, AND SCHMIDT

Appell, David. "Dark Forces at Work." *Scientific American.* May 1, 2008. Accessed August 2015. http://www.scientificamerican.com/article/dark-forces-at-work/?page=2.

"The High-Z SN Search." Harvard-Smithsonian Center for Astrophysics. Accessed August 2015. https://www.cfa.harvard.edu/supernova//HighZ.html.

Panek, Richard. *The 4% Universe: Dark Matter, Dark Energy, and the Race to Discover the Rest of Reality.* Boston: Houghton Mifflin Harcourt, 2011.

Preuss, Paul. "The Evolving Search for the Nature of Dark Energy." Berkeley Lab, October 27, 2009. Accessed August 2015. http://newscenter.lbl.gov/2009/10/27/evolving-dark-energy/.

"Saul Perlmutter." *Wikipedia*, s.v. Last modified October 1, 2015. https://en.wikipedia.org/wiki/Saul_Perlmutter.

CHAPTER 15.3: TEN HONORABLE MENTION: PARTICLE ACCELERATIONS

"The History of CERN." CERN Timelines. Accessed August 2015. http://timeline.web.cern.ch/timelines/The-history-of-CERN.

Kolbert, Elizabeth. "Crash Course: Can a Seventeen-Mile-Long Collider Unlock the Universe?" *New Yorker*, May 14, 2007, Accessed August 2015. http://www.newyorker.com/magazine/2007/05/14/crash-course.

CHAPTER 15.4: TEN HONORABLE MENTION: BOSE, BOSONS, AND THE HIGGS BOSON

"Satyendranath Bose." MacTutor History of Mathematics Archive. Accessed August 2015. http://www-history.mcs.st-and.ac.uk/Biographies/Bose.html.

"Satyendra Nath Bose." *Wikipedia*, s.v. Last modified January 3, 2016. https://en.wikipedia.org/wiki/Satyendra_Nath_Bose.

CHAPTER 15.5: TEN HONORABLE MENTION: ENRICO FERMI, FRANK DRAKE, AND JILL TARTER

"Breakthrough Initiatives." *Wikipedia*, s.v. Last modified October 20, 2015. https://en.wikipedia.org/wiki/Breakthrough_Initiatives.

"*Contact* (1997 American Film)." *Wikipedia*, s.v. Last modified December 3, 2015. https://en.wikipedia.org/wiki/Contact_(1997_American_film).

"Drake Equation." *Wikipedia*, s.v. Last modified, January 4, 2016. https://en.wikipedia.org/wiki/Drake_equation.

"Jill Tarter." *Wikipedia*, s.v. Last modified December 11, 2015. https://en.wikipedia.org/wiki/Jill_Tarter.

"Jill Tarter: A Scientist Searching for Alien Life." NPR, July 23, 2012. Accessed August 2015. http://www.npr.org/2012/07/23/156366055/jill-tarter-a-scientist-searching-for-alien-life.

Shostak, Seth. "Should We Keep a Low Profile in Space?" *New York Times*, March 25, 2015. Accessed August 2015. http://www.nytimes.com/2015/03/28/opinion/sunday/messaging-the-stars.html.

"Yuri Milner." *Wikipedia*, s.v. Last modified December 24, 2015. https://en.wikipedia.org/wiki/Yuri_Milner.

CHAPTER 15.6: TEN HONORABLE MENTION: STEPHEN HAWKING AND BLACK HOLES

Ferguson, Kitty. *Stephen Hawking: An Unfettered Mind*. New York: Macmillan, 2012.

Mlodinow, Leonard. *The Upright Thinkers: The Human Journey from Living in Trees to Understanding the Cosmos*. New York: Pantheon, 2015.

"Stephen Hawking." *Wikipedia*, s.v. Last modified January 6, 2016. https://en.wikipedia.org/wiki/Stephen_Hawking.

Susskind, Leonard. *The Black Hole War: My Battle with Stephen Hawking to Make the World Safe for Quantum Mechanics*. New York: Little, Brown, 2008.

CHAPTER 15.7: TEN HONORABLE MENTION: COSMOS

Achenbach, Joel. "Why Carl Sagan Is Truly Irreplaceable." *Smithsonian*, March 2014. Accessed August 2015. http://www.smithsonianmag.com/science-nature/why-carl-sagan-truly-irreplaceable-180949818/.

"Carl Sagan." *Wikipedia*, s.v. Last modified January 4, 2016. https://en.wikipedia.org/wiki/Carl_Sagan.

"*Cosmos: A Spacetime Odyssey.*" *Wikipedia*, s.v. Last modified December 30, 2015. https://en.wikipedia.org/wiki/Cosmos:_A_Spacetime_Odyssey.

Morrison, David. "Carl Sagan's Life and Legacy as Scientist, Teacher, and Skeptic." *Skeptical Inquirer* 31, no. 1 (January/February 2007). Accessed August 2015. http://www.csicop.org/si/show/carl_sagans_life_and_legacy_as_scientist_teacher_and_skeptic.

Sagan, Carl, and Ann Druyan. *The Demon-Haunted World: Science as a Candle in the Dark*. New York: Ballantine, 1996.

CHAPTER 15.8: TEN HONORABLE MENTION: GEORGE WASHINGTON CARVER

Bolden, Tanya. *George Washington Carver*. New York: Harry N. Abrams, 2008.

"George Washington Carver." *Wikipedia*, s.v. Last modified January 4, 2016. https://en.wikipedia.org/wiki/George_Washington_Carver.

Wormser, Richard. "The Rise and Fall of Jim Crow." PBS.org. Accessed August 2015. http://www.pbs.org/wnet/jimcrow/stories_people_booker.html.

CHAPTER 15.9: TEN HONORABLE MENTION: HEDY LAMARR AND GEORGE ANTHÉIL

"George Antheil." *Wikipedia*, s.v. Last modified January 6, 2016. https://en.wikipedia.org/wiki/George_Antheil.

"Hedy Lamarr." *Wikipedia*, s.v. Last modified January 6, 2016. https://en.wikipedia.org/wiki/Hedy_Lamarr.

"Hedy Lamarr: Invention of Spread Spectrum Technology." Famous Women Inventors. Accessed August 2015. http://www.women-inventors.com/Hedy-Lammar.asp.

Rhodes, Richard. *Hedy's Folly: The Life and Breakthrough Inventions of Hedy Lamarr, the Most Beautiful Woman in the World*. New York: Vintage, 2011.

Shearer, Stephen M. *Beautiful: The Life of Hedy Lamarr*. New York: St. Martin's Griffin, 2013.

CHAPTER 15.10: TEN HONORABLE MENTION: THE AMAZING RANDI

"Billet Reading." *Wikipedia*, s.v. Last modified November 24, 2015. https://en.wikipedia.org/wiki/Billet_reading.

Higginbotham, Adam. "The Unbelievable Skepticism of the Amazing Randi." *New York Times Magazine*, November 7, 2014. Accessed August 2015. http://www.nytimes.com/2014/11/09/magazine/the-unbelievable-skepticism-of-the-amazing-randi.html?r=0.

James Randi Educational Foundation. Accessed August 2015. http://web.randi.org/about.html.

INDEX